Design of Experiments for Enginee~ ~d S

Design of Experiments for Engineers and Scientists

Jiju Antony

AMSTERDAM BOSTON HEIDELBERG LONDON NEW YORK OXFORD
PARIS SAN DIEGO SAN FRANCISCO SINGAPORE SYDNEY TOKYO

Butterworth-Heinemann is an imprint of Elsevier
Linacre House, Jordan Hill, Oxford OX2 8DP, UK
30 Corporate Drive, Suite 400, Burlington, MA 01803, USA

First edition 2003
Reprinted 2004, 2005, 2006, 2007 (twice), 2008

British Library Cataloguing in Publication Data
A catalogue record for this book is available from the British Library

Library of Congress Cataloging-in-Publication Data
A catalog record for this book is available from the Library of Congress

ISBN: 978- 0-7506-4709-0

For information on all Butterworth-Heinemann publications
visit our website at www.elsevierdirect.com

Transferred to Digital Printing in 2010

This book is dedicated to my late papa who instilled in me the importance of education

Contents

Preface

Design of Experiments (DOE) is a powerful technique used for exploring new processes, gaining increased knowledge of the existing processes and optimizing these processes for achieving world class performance. The author's involvement in promoting and training the use of DOE dates back to mid-1990s. There are plenty of books available in the market today on this subject written by classic statisticians though majority of them are suited to other statisticians than to run-of-the-mill industrial engineers and business managers with limited mathematical and statistical skills.

DOE never has been a favourite technique for many of today's engineers and managers in organizations due to the number crunching involved and the statistical jargon incorporated into the teaching mode by many statisticians. This book is targeted for people who have either been intimidated by their attempts to learn about DOE or never appreciated the true potential of DOE for achieving breakthrough improvements in product quality and process efficiency.

This book gives a solid introduction to the technique through a myriad of practical examples and case studies. The readers of this book will develop a sound understanding of the theory of DOE and practical aspects of how to design, analyse and interpret the results of a designed experiment. Throughout this book, the emphasis is on the simple but powerful graphical tools for data analysis and interpretation. All of the graphs and figures in this book were created using Minitab version 13.0 for Windows.

The author sincerely hopes that practising industrial engineers and managers as well as researchers in academic world will find this book useful in learning how to apply DOE in their own work environment. The book will also be a useful resource for people involved in Six Sigma training and projects related to design optimization and process performance improvements.

The author hopes that this book inspires readers to get into the habit of applying DOE for problem solving and process trouble-shooting. The author strongly recommends readers of this book to continue on a more advanced reference to learn about topics which are not covered here. The author is indebted to many contributors and gurus to the development of various experimental design techniques, especially Sir Ronald Fisher, Plackett and Burman, Professor George Box, Professor Douglas Montgomery, Dr Genichi Taguchi and Dr Dorian Shainin.

Acknowledgements

This book was conceived further to my publication of an article entitled 'Teaching Experimental Design techniques to Engineers and Managers' in the *International Journal of Engineering Education*. I am deeply indebted to a number of people who in essence, have made this book what it is today. First, and foremost, I would like to thank Dr Hefin Rowlands, Head of Research and Enterprise of the University of Wales, Newport, for his constructive comments on the earlier drafts of the chapters. I am also indebted to the quality and production managers of companies that I have been privileged to work with and gather data. I would also like to take this opportunity to thank my students both on-campus and off-campus.

I would like to express my deepest appreciation to Claire Harvey, the commissioning editor of Elsevier Science, for her incessant support and forbearance, during the course of this project. Finally, I express my sincere thanks to my wife, Frenie and daughter Evelyn, for their encouragement and patience as the book stole countless hours away from family activities.

1

Introduction to industrial experimentation

1.1 Introduction

Experiments are performed today in many manufacturing organizations to increase our understanding and knowledge of various manufacturing processes. Experiments in manufacturing companies are often conducted in a series of trials or tests which produce quantifiable outcomes. For continuous improvement in product/process quality, it is fundamental to understand the process behaviour, the amount of variability and its impact on processes. In an engineering environment, experiments are often conducted to explore, estimate or confirm. Exploration refers to understanding the data from the process. Estimation refers to determining the effects of process variables or factors on the output performance characteristic. Confirmation implies verifying the predicted results obtained from the experiment.

In manufacturing processes, it is often of primary interest to explore the relationships between the key input process variables (or factors) and the output performance characteristics (or quality characteristics). For example, in a metal cutting operation, cutting speed, feed rate, type of coolant, depth of cut, etc. can be treated as input variables and surface finish of the finished part can be considered as an output performance characteristic.

One of the common approaches employed by many engineers today in manufacturing companies is One-Variable-At-a-Time (OVAT), where we vary one variable at a time keeping all other variables in the experiment fixed. This approach depends upon guesswork, luck, experience and intuition for its success. Moreover, this type of experimentation requires large resources to obtain a limited amount of information about the process. One Variable-At-a-Time experiments often are unreliable, inefficient, time consuming and may yield false optimum condition for the process.

Statistical thinking and statistical methods play an important role in planning, conducting, analysing and interpreting data from engineering experiments. When several variables influence a certain characteristic of a product, the best strategy is then to design an experiment so that valid, reliable and sound conclusions can be drawn effectively, efficiently and economically.

In a designed experiment, the engineer often makes deliberate changes in the input variables (or factors) and then determines how the output functional performance varies accordingly. It is important to note that not all variables affect the performance in the same manner. Some may have strong influences on the output performance, some may have medium influences and some have no influence at all. Therefore, the objective of a carefully planned designed experiment is to understand which set of variables in a process affects the performance most and then determine the best levels for these variables to obtain satisfactory output functional performance in products.

Design of Experiments (DOE) was developed in the early 1920s by Sir Ronald Fisher at the Rothamsted Agricultural Field Research Station in London, England. His initial experiments were concerned with determining the effect of various fertilizers on different plots of land. The final condition of the crop was not only dependent on the fertilizer but also on a number of other factors (such as underlying soil condition, moisture content of the soil, etc.) of each of the respective plots. Fisher used DOE which could differentiate the effect of fertilizer and the effect of other factors. Since then DOE has been widely accepted and applied in biological and agricultural fields. A number of successful applications of DOE have been reported by many US and European manufacturers over the last fifteen years or so. The potential applications of DOE in manufacturing processes include:

- improved process yield and stability
- improved profits and return on investment
- improved process capability
- reduced process variability and hence better product performance consistency
- reduced manufacturing costs
- reduced process design and development time
- heightened morale of engineers with success in chronic-problem solving
- increased understanding of the relationship between key process inputs and output(s)
- increased business profitability by reducing scrap rate, defect rate, rework, retest, etc.

Industrial experiments involves a sequence of activities:

1. *Hypothesis* – an assumption that motivates the experiment
2. *Experiment* – a series of tests conducted to investigate the hypothesis
3. *Analysis* – involves understanding the nature of data and performing statistical analysis of the data collected from the experiment
4. *Interpretation* – is about understanding the results of the experimental analysis
5. *Conclusion* – involves whether or not the originally set hypothesis is true or false. Very often more experiments are to be performed to test the hypothesis and sometimes we establish new hypothesis which requires more experiments.

Consider a welding process where the primary concern of interest to engineers is the strength of the weld and the variation in the weld strength values. Through scientific experimentation, we can determine what factors mostly affect the mean weld strength and variation in weld strength. Through experimentation, one can also predict the weld strength under various conditions of key input welding machine parameters or factors (e.g. weld speed, voltage, welding time, weld position, etc.).

For the successful application of an industrial designed experiment, we generally require the following skills:

- *Planning skills* Understanding the significance of experimentation for a particular problem, time and budget required for the experiment, how many people are involved with the experimentation, establishing who is doing what, etc.
- *Statistical skills* Involve the statistical analysis of data obtained from the experiment, assignment of factors and interactions to various columns of the design matrix (or experimental layout), interpretation of results from the experiment for making sound and valid decisions for improvement, etc.
- *Teamwork skills* Involve understanding the objectives of the experiment and having a shared understanding of the experimental goals to be achieved, better communication among people with different skills and learning from one another, brainstorming of factors for the experiment by team members, etc.
- *Engineering skills* Determination of the number of levels of each factor, range at which each factor can be varied, determination of what to measure within the experiment, determination of capability of the measurement system in place, determination of what factors can be controlled and what cannot be controlled for the experiment, etc.

1.2 Some fundamental and practical issues in industrial experimentation

An engineer is interested in measuring the yield of a chemical process, which is influenced by two key process variables (or control factors). The engineer decides to perform an experiment to study the effects of these two variables on the process yield. The engineer uses an OVAT approach to experimentation. The first step is to keep the temperature constant (T_1) and vary the pressure from P_1 to P_2. The experiment is repeated twice and the results are illustrated in Table 1.1. The engineer conducts four experimental trials.

The next step is to keep the pressure constant (P_1) and vary the temperature from T_1 to T_2. The results of the experiment are shown in Table 1.2.

The engineer has calculated the average yield values for only three combinations of temperature and pressure: (T_1, P_1), (T_1, P_2) and (T_2, P_1).

Table 1.1 The effects of varying pressure on process yield

Trial	Temperature	Pressure	Yield	Average yield (%)
1	T_1	P_1	55, 57	56
2	T_1	P_2	63, 65	64

Table 1.2 The effects of varying pressure on process yield

Trial	Temperature	Pressure	Yield	Average yield (%)
3	T_1	P_1	55, 57	56
4	T_2	P_1	60, 62	61

The engineer concludes from the experiment that the maximum yield of the process can be attained by corresponding to (T_1, P_1). The question then arises as to what should be the average yield corresponding to the combination (T_2, P_2)? The engineer was unable to study this combination as well as the interaction between temperature and pressure. Interaction between two factors exists when the effect of one factor on the response or output is different at different levels of the other factor. The difference in the average yield between the trials one and two provides an estimate of the effect of pressure. Similarly, the difference in the average yield between trials three and four provides an estimate of the effect of temperature. An effect of a factor is the change in the average response due to a change in the levels of a factor. The effect of pressure was estimated to be 8 per cent (i.e. $64 - 56$) when temperature was kept constant at 'T_1'. There is no guarantee whatsoever that the effect of pressure will be the same when the conditions of temperature change. Similarly the effect of temperature was estimated to be 5 per cent (i.e. $61 - 56$) when pressure was kept constant at 'P_1'. It is reasonable to say that we do not get the same effect of temperature when the conditions of pressure change. Therefore the OVAT approach to experimentation can be misleading and may lead to unsatisfactory experimental conclusions in real life situations. Moreover, the success of OVAT approach to experimentation relies on guesswork, luck, experience and intuition. This type of experimentation is inefficient in that it requires large resources to obtain a limited amount of information about the process. In order to obtain a reliable and predictable estimate of factor effects, it is important that we should vary the factors simultaneously at their respective levels. In the above example, the engineer should have varied the levels of temperature and pressure simultaneously to obtain reliable estimates of the effects of temperature and pressure. Experiments of this type will be the focus of the book.

1.3 Summary

This chapter illustrates the importance of experimentation in organizations and a sequence of activities to be taken into account while performing an industrial experiment. The chapter briefly illustrates the key skills required for the successful application of an industrial designed experiment. The fundamental problems associated with OVAT approach to experimentation are also demonstrated in the chapter with an example.

Exercises

1. Why do we need to perform experiments in organizations?
2. What are the limitations of OVAT approach to experimentation?
3. What factors make an experiment successful in organizations?

References

Antony, J. (1997). A Strategic Methodology for the Use of Advanced Statistical Quality Improvement Techniques, *PhD Thesis*, University of Portsmouth, UK.
Clements, R.B. (1995). *The Experimenter's Companion*. Wisconsin, USA, ASQC Quality Press.
Montgomery, D.C. et al. (1998). *Engineering Statistics*. USA, John Wiley and Sons.

2

Fundamentals of Design of Experiments

2.1 Introduction

In order to properly understand a designed experiment, it is essential to have a good understanding of the process. A process is the transformation of inputs into outputs. In the context of manufacturing, inputs are factors or process variables such as people, materials, methods, environment, machines, procedures, etc. and outputs can be performance characteristics or quality characteristics of a product. Sometimes, an output can also be referred to as response.

In performing a designed experiment, we will intentionally make changes to the input process or machine variables (or factors) in order to observe corresponding changes in the output process. The information gained from properly planned, executed and analysed experiments can be used to improve functional performance of products, to reduce scrap rate or rework rate, to reduce product development cycle time, to reduce excessive variability in production processes, etc. Let us suppose that an experimenter wishes to study the influence of six variables or factors on an injection moulding process. Figure 2.1 illustrates an example of an injection moulding process with possible inputs and outputs.

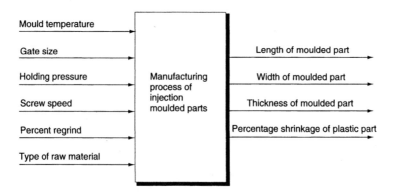

Figure 2.1 Illustration of an injection moulding process.

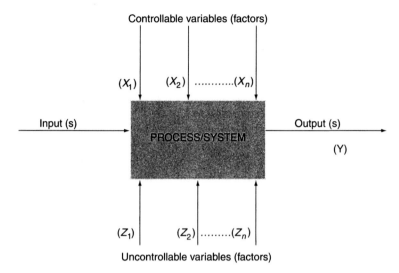

Figure 2.2 General model of a process/system.

In real life situations, some of the process variables or factors can be controlled fairly easily and some of them are hard or expensive to control during normal production or standard conditions. Figure 2.2 illustrates a general model of a process or system.

In the above diagram, output(s) are performance characteristics which are measured to assess process/product performance. Controllable variables (represented by X's) can be varied easily during an experiment and such variables have a key role to play in the process characterization. Uncontrollable variables (represented by Z's) are difficult to control during an experiment. These variables or factors are responsible for variability in product performance or product performance inconsistency. It is important to determine the optimal settings of X's in order to minimize the effects of Z's. This is the fundamental strategy of robust design.

2.2 Basic principles of Design of Experiments

Design of Experiments refers to the process of planning, designing and analysing the experiment so that valid and objective conclusions can be drawn effectively and efficiently. In order to draw statistically sound conclusions from the experiment, it is necessary to integrate simple and powerful statistical methods into the experimental design methodology. The success of any industrially designed experiment depends on sound planning, appropriate choice of design, statistical analysis of data and teamwork skills.

In the context of DOE in manufacturing, one may come across two types of process variables or factors: qualitative and quantitative factors. For quantitative factors, one must decide on the range of settings, how they are to be measured and controlled during the experiment. For example, in the above injection moulding process, screw speed, mould temperature, etc. are examples of quantitative factors. Qualitative factors are discrete in nature. Type of raw material, type of catalyst, type of supplier, etc. are examples of qualitative factors. A factor may take different levels, depending on the nature of the factor – quantitative or qualitative. A qualitative factor generally requires more levels when compared to a quantitative factor. Here the term 'level' refers to a specified value or setting of the factor being examined in the experiment. For instance, if the experiment is to be performed using three different types of raw materials, then we can say that the factor, type of raw material, has three levels. In the DOE terminology, a trial or run is a certain combination of factor levels whose effect on the output (or performance characteristic) is of interest.

The three principles of experimental design such as randomization, replication and blocking can be utilized in industrial experiments to improve the efficiency of experimentation. These principles of experimental design are applied to reduce or even remove experimental bias. It is important to note that large experimental bias could result in wrong optimal settings or in some cases it could mask the effect of the really significant factors. Thus an opportunity for gaining process understanding is lost, and a primary element for process improvement is overlooked.

2.2.1 Randomization

We all live in a non-stationary world, a world in which noise factors (or external disturbances) never stay still. For instance, the manufacture of a metal part is an operation involving people, machines, measurement, environment, etc. The parts of the machine are not fixed entities; they are wearing out over a period of time and their accuracy is not constant over a period of time. The attitude of the people who operate the machines varies from time to time. If you believe your system or process is stable, you do not then need to randomize the experimental trials. On the other hand, if you believe your process is unstable and without randomization the results would be meaningless and misleading, you do then need to think about randomization of experimental trials. If the process is very unstable and randomization would make your experiment impossible, then do not run the experiment. You may have to look at process control methods to bring your process into a state of statistical control.

While designing industrial experiments, there are factors, such as power surges, operator errors, fluctuations in ambient temperature and humidity, raw material variations, etc. which may influence the process output performance because they are often expensive or difficult to control. Such factors can adversely affect the experimental results and therefore must be either minimized or removed from the experiment. Randomization is one of the methods experimenters often rely on to reduce the effect of experimental bias. By properly randomizing the experiment, we assist in averaging out the effects of noise factors that may be present in the process. In other words, randomization can ensure that all levels of a factor have an equal chance of being affected by noise factors. Dorian Shainin accentuates the importance of randomization as 'experimenters' insurance policy'. He pointed out that 'failure to randomize the trial conditions mitigates the statistical validity of an experiment'.

Sometimes experimenters encounter situations where randomization of experimental trials is difficult to perform due to cost and time constraints. For instance, temperature in a chemical process may be a hard-to-change factor, making complete randomization of this factor almost impossible. Under such circumstances, it might be desirable to change the factor levels of temperature less frequently than others. In such situations, *restricted randomization* can be employed.

It is important to note that in classical DOE approach, complete randomization of the experimental trials is advocated whereas in Taguchi approach to experimentation, the incorporation of noise factors into the experimental layout will supersede the need for randomization. The following tips are useful if you decide to apply randomization strategy to your experiment.

- What is the cost associated with change of factor levels?
- Have we incorporated any noise factors in the experimental layout?
- What is the set up time between trials?
- How many factors in the experiment are expensive or difficult to control?
- Where do we assign factors whose levels are difficult to change from one to another level?

2.2.2 Replication

Replication is a process of running the experimental trials in a random sequence. Replication means repetitions of an entire experiment or a portion of it, under more than one condition. Replication has two important properties. The first property is that it allows the experimenter to obtain an estimate of the experimental error. The second property is that it permits the experimenter to obtain a more precise estimate of the factor/interaction effect. If the number of replicates is equal to one or unity, we would not then be able to make satisfactory conclusions about the effect of either factors or interactions. The factor or interaction effect could be significant due to the result of experimental error. On the other hand, if we have sufficient number of replicates, we would be safely making satisfactory inferences about the effect of factors/interactions.

Replication can result in a substantial increase in the time to conduct an experiment. Moreover, if the material is expensive, replication may lead to exorbitant material costs. Any bias or experimental error associated with set-up changes will be evenly distributed across the experimental runs or trials using replication. The use of replication in real life must be justified in terms of time and cost.

Many experimenters use the terms 'repetition' and 'replication' inter-changeably. Technically speaking, they are not the same. In repetition, an experimenter may repeat an experimental trial condition a number of times as planned, before proceeding to the next trial in the experimental layout. The advantage of this approach is that the experimental set-up cost should be minimum. However, a set-up error is unlikely to be detected or identified.

2.2.3 Blocking

Blocking is a method of eliminating the effects of extraneous variation due to noise factors and thereby improves the efficiency of experimental design. The main objective is to eliminate unwanted sources of variability such as batch-to-batch, day-to-day, shift-to-shift, etc. The idea is to arrange similar experimental runs into blocks (or groups). Generally, a block is a set of relatively homogeneous experimental conditions. The blocks can be batches of raw materials, different operators, different vendors, etc. Observations collected under the same experimental conditions (i.e. same day, same shift, etc.) are said to be in the same block. Variability between blocks must be eliminated from the experimental error, which leads to an increase in the precision of the experiment. The following two examples illustrate the role of blocking in industrial designed experiments.

> *Example 2.1*
> A metallurgist wants to improve the strength of a steel product Four factors are being considered for the experiment, which might have some impact on the strength. It is decided to study each factor at 2-levels (i.e. a low setting and a high setting). An eight trial experiment is chosen by the experimenter but only four trials are possible to run per day. Here each day can be treated as a separate block.

> *Example 2.2*
> An experiment in a chemical process requires two batches of raw material for conducting the entire experimental runs. In order to minimize the effect of batch-to-batch material variability, we need to treat batch of raw material as a noise factor. In other words, each batch of raw material would form a block.

2.3 Degrees of freedom

In the context of statistics, the term 'degrees of freedom' is the number of independent and fair comparisons that can be made in a set of data. For example, consider the height of two students, say John and Kevin. If the height of John is H_J and that of Kevin is H_K, then we can make only one fair comparison $(H_J - H_K)$.

In the context of DOE, the number of degrees of freedom associated with a process variable is equal to one less than the number of levels for that factor. For example, an engineer wishes to study the effects of reaction temperature and reaction time on the yield of a chemical process. Assume each factor was studied at two levels. The number of degrees of freedom associated with each factor is equal to unity or one (i.e. $2 - 1 = 1$).

Degrees of freedom for a main effect = Number of levels − 1

The number of degrees of freedom for the entire experiment is equal to one less than the total number of data points or observations. Assume that you have performed an eight trial experiment and each trial condition was replicated twice. The total number of observations in this case is equal to 16 and therefore the total degrees of freedom for the experiment is equal to 15 (i.e. $16 - 1$).

The degrees of freedom for an interaction is equal to the product of the degrees of freedom associated with each factor involved in that particular interaction effect. For instance, in the above yield example, the degrees of freedom for both reaction temperature and reaction time is equal to one and therefore, the degrees of freedom for its interaction effect is also equal to unity.

2.4 Confounding

The term confounding refers to the combining influences of two or more factor effects in one measured effect. In other words, one cannot estimate factor effects and their interaction effects independently. Effects which are confounded are called aliases. A list of the confoundings which occur in an experimental design is called an alias structure or a confounding pattern. The confounding of effects is simple to illustrate. Suppose two factors, say, mould temperature and injection speed are investigated at 2-levels. Five response values are taken when both factors are at their low levels and high levels respectively. The results of the experiment (i.e. mean response) are as shown in Table 2.1.

The effect of mould temperature is equal to $82.75 - 75.67 = 7.08$. Here effect refers to the change in mean response due to a change in the levels of a factor.

Table 2.1 Example of confounding

Mould temperature	Injection speed	Mean response
Low level	Low level	75.67
High level	High level	82.75

The effect of injection speed is also same as that of mould temperature (i.e. 82.75 − 75.67). So is the calculated effect actually due to injection speed or mould temperature? One cannot simply tell this as the effects are confounded.

2.5 Design resolution

Design resolution (R) is a summary characteristic of aliasing or confounding patterns. The degree to which the main effects are aliased with the interaction effects (two-factor or higher) is represented by the resolution of the corresponding design. Obviously, we don't prefer the main effects to be aliased with other main effects. A design is of resolution R if no p-factor effect is aliased with another effect containing less than (R − p) factors. For designed experiments, designs of resolution III, IV and V are particularly important.

Design resolution identifies for a specific design, the order of confounding of the main effects and their interactions. It is a key tool for determining what fractional factorial design will be the best choice for a given problem. More information on full and fractional factorial designs can be seen in later chapters of this book.

Resolution III designs These are designs in which no main effects are confounded with any other main effect, but main effects are confounded with two-factor interactions and two-factor interactions may be confounded with each other.

Resolution IV designs These are designs in which no main effects are confounded with any other main effect or with any two-factor interaction effects, but two-factor interaction effects are confounded with each other.

Resolution V designs These are designs in which main effects are not confounded with other main effects, two-factor interactions or three-factor interactions; but two-factor interactions are confounded with three-factor interactions.

2.6 Metrology considerations for industrial designed experiments

For industrial experiments, the response or quality characteristic will have to be measured either by direct or by indirect methods. These measurement methods produce variation in the response. Measurement is a process, and varies, just as all processes vary. Identifying, separating and removing the measurement variation lead to improvements to the actual measured values obtained from the use of the measurement process.

The following characteristics need to be considered for a measurement system:

- *Accuracy* It refers to the degree of closeness between the measured value and the true value or reference value.

- *Precision*   It is a measure of the scatter of results of several observations and is not related to the true value. It is a comparative measure of the observed values and is only a measure of the random errors. It is expressed quantitatively as the standard deviation of observed values from repeated results under identical conditions.
- *Stability*   A measurement system is said to be stable if the measurements do not change over time. In other words, they should not be adversely influenced by operator and environmental changes.
- *Capability*   A measurement system is capable if the measurements are free from bias (accurate) and sensitive. A capable measurement system requires sensitivity (the variation around the average should be small compared to the specification limits or process spread and accuracy).

2.6.1 Measurement system capability

The goal of a measurement system capability study is to understand and quantify the sources of variability present in the measurement system. Repeatability and Reproducibility (R&R) studies analyse the variation of measurements of a gauge and the variation of measurements by operators respectively. Repeatability refers to the variation in measurements obtained when an operator uses the same gauge several times for measuring the identical characteristic on the same part. Reproducibility, on the other hand, refers to the variation in measurements when several operators use the same gauge for measuring the identical characteristic on the same part. It is important to note that total variability in a process can be broken down into variability due to product (or parts variability) and variability due to measurement system. The variability due to measurement system is further broken into variability due to gauge (i.e. repeatability) and reproducibility. Reproducibility can be further broken into variability due to operators and variability due to (part × operator) interaction.

A measurement system is considered to be capable and adequate if it satisfies the following criterion:

$$\frac{P}{T} \leq 10\% \tag{2.1}$$

where P/T = Precision-to-Tolerance ratio, which is given by:

$$\frac{P}{T} = \frac{6\hat{\sigma}_{\text{measurement error}}}{\text{USL} - \text{LSL}} \tag{2.2}$$

Moreover,

$$\hat{\sigma}^2_{\text{measurement error}} = \hat{\sigma}^2_{\text{repeatability}} + \hat{\sigma}^2_{\text{reproducibility}}$$

There are obvious dangers in relying too much on the P/T ratio. For example, the P/T ratio may be made arbitrarily small by increasing the width of the specification of tolerance band. The gauge must be able to have sufficient capability to detect meaningful variation in the product. The contribution of gauge variability (or measurement error) to the total variability is a much more useful criterion for determining the measurement system capability. So one may look at the following equation as to see whether the given measurement system is capable or not.

$$\frac{\hat{\sigma}_{\text{measurement error}}}{\hat{\sigma}_{\text{total}}} \leq 10\% \tag{2.3}$$

Another useful gauge to evaluate a measurement system is to see whether or not the measurement process is able to detect product variation. If the amount of measurement system variability is high, it will obscure the product variation. It is important to be able to separate out measurement variability from product variability. Donald J. Wheeler uses discrimination ratio as an indicator of whether the measurement process is able to detect product variation. For more information on discrimination ratio and its use in gauge capability analysis, I would advice the readers to refer to his book entitled 'Evaluating the Measurement Process' (see reference list).

2.6.2 Some tips for the development of a measurement system

The key to managing processes is measurement. Engineers and managers, therefore, must strive to develop useful measurements of their processes. The following tips are useful when developing a measurement system for industrial experiments.

1. *Select the process you want to measure.* This involves process definition and determination of recipients of the information on measurements, and how it will be used.
2. *Define the characteristic that needs to be measured within the process.* This involves identification and definition of suitable characteristics that reflect customer needs and expectations. It is always best to have a team of people comprising members from quality engineering, process engineering and operators in defining the key characteristics that need to be measured within a process.
3. *Perform a quality check.* It is quite important to address the following questions during the development of a measurement system.

   (a) How accurately can we measure the product characteristics?
   (b) What is the error in our measurement system? Is it acceptable?
   (c) Is our measurement system stable and capable?
   (d) What is the contribution of our measurement system variability to the total variation? Is it acceptable?

2.7 Selection of quality characteristics for industrial experiments

The selection of an appropriate quality characteristic is vital for the success of an industrial experiment. To identify a good quality characteristic, it is suggested to start with the engineering or economic goal. Having determined this goal, identify the fundamental mechanisms and the physical laws affecting this goal. Finally, choose the quality characteristics to increase the understanding of these mechanisms and physical laws. The following points are useful in selecting the quality characteristics for industrial experiments:

- Try to use quality characteristics which are easy to measure.
- Quality characteristics should be continuous variables as far as possible.
- Use quality characteristics which can be measured precisely, accurately and with stability.
- For complex processes, it is best to select quality characteristics at the sub-system level and perform experiments at this level prior to attempting overall process optimization.
- Quality characteristics should cover all dimensions of the ideal function or the input-output relationship.
- Quality characteristics should preferably be additive (i.e. no interaction exists among the quality characteristics) and monotonic (i.e. the effect of each factor on robustness should be in a consistent direction, even when the settings of factors are changed).

Consider a certain painting process which results in various problems such as orange peel, poor appearance, voids, etc. Too often, experimenters measure these characteristics as data and try to optimize the quality characteristic. It is not the function of the coating process to produce an orange peel. The problem could be due to excess variability of the coating process due to noise factors such as variability in viscosity, ambient temperature, etc. We should make every effort to gather data that relate to the engineering function itself and not to the symptom of variability. One fairly good characteristic to measure for the coating process is the coating thickness. It is important to understand that excess variability of coating thickness from its target value could lead to problems such as orange peel or voids. The sound engineering strategy is to design and analyse an experiment so that best process parameter settings can be determined which yields minimum variability of coating thickness around the specified target thickness.

Exercises

1. What are the three basic principles of DOE?
2. Explain the role of randomization in industrial experiments. What are the limitations of randomization in experiments?

3. What is replication? Why do we need to replicate experimental trials?
4. What is the fundamental difference between repetition and replication?
5. Explain the term degrees of freedom?
6. What is confounding and what is its role in the selection of a particular design matrix or experimental layout?
7. What is design resolution and briefly illustrate its significance in industrial experiments?
8. What is the role of measurement system in the context of industrial experimentation?
9. State three key factors for the selection of quality characteristics for the success of an industrial experiment.

References

Antony, J. (1997). A Strategic Methodology for the Use of Advanced Statistical Quality Improvement Techniques, *PhD Thesis*, University of Portsmouth, UK.

Antony, J. (1998). Some key Things Industrial Engineers Should Know about Experimental Design. *Logistics Information Management*, 11(6), 386–392.

Barker, T.B. (1990). *Engineering Quality by Design-Interpreting the Taguchi Approach*. USA, Marcel Dekker Inc.

Belavendram, N. (1995). *Quality by Design: Taguchi Techniques for Industrial Experimentation*. UK, Prentice-Hall Publishers.

Bisgaard, S. (1994). Blocking Generators for Small $2^{(k-p)}$ designs. *Journal of Quality Technology*, pp. 288–296.

Box, G.E.P. (1990). Must we Randomise our Experiment? *Quality Engineering*, 2(4), 497–502.

Kolarik, W.J. (1995). *Creating Quality: Concepts, Systems, Strategies and Tools*. USA, McGraw-Hill.

Leon, R.V., Shoemaker, A. and Tsui, K.L. (1993). Discussion on Planning for a Designed Industrial Experiment. *Technometrics*, 35(1), 21–24.

Mitra, A. (1993). *Fundamentals of Quality Control and Improvement*. USA, Macmillan Publishers.

Montgomery, D.C. and Runger, G.C. (1993/94). Gauge Capability and Designed Experiments – Part 1: Basic methods. *Quality Engineering*, 6(1), 115–135.

Montgomery, D.C. (2001). *Design and Analysis of Experiments*. USA, John Wiley and Sons.

Roy, K. (2001). *Design of Experiments using the Taguchi Approach*. USA, John Wiley and Sons.

Vecchio, R.J. (1997). *Understanding Design of Experiments*. USA, Gardner Publications.

Wheeler, D.J. and Lyday, R.W. (1989). *Evaluating the Measurement Process*. USA, SPC Press.

3

Understanding key interactions in processes

3.1 Introduction

For modern industrial processes, the interactions between the factors or process parameters are a major concern to many engineers and managers, and therefore should be studied, analysed and understood properly for problem solving and process optimization problems. For many process optimization problems in industries, the root cause of the problem is sometimes the interaction between the factors rather than the individual effect of each factor on the output performance characteristic (or response). Here performance characteristic is the characteristic of a product/service which is most critical to customers.

The significance of interactions in manufacturing processes can be illustrated by the following example taken from a wave-soldering process of a PCB assembly line in a certain electronic company. The engineering team of the company was interested in reducing the number of defective solder joints obtained from the soldering process. The average defect rate based on the existing conditions is 410 ppm (parts-per-million). The team has decided to perform a simple experiment to understand the influence of wave-soldering process parameters on the number of defective solder joints.

The team initially utilised an OVAT approach to experimentation. Each process parameter (or process variable) was studied at two levels – low level (represented by -1) and high level (represented by $+1$). The parameters and their levels are shown in Table 3.1. The experimental layout (or design matrix) for this experiment is shown in Table 3.2. The design matrix shows all the possible combinations of factors at their respective levels.

In the experimental layout, the actual process parameter settings are replaced by -1 and $+1$. The first trial in Table 3.2 represents the current process settings, with each process parameter kept at low level. In the second trial, the team has changed the level of factor 'A' from low to high, keeping the levels of other two factors constant. The engineer notices from this experiment that the defect rate is minimum corresponding to trial condition 4, and thereby concludes that the optimal setting is the one corresponding to fourth trial.

Table 3.1 List of process parameters and their levels

Labels	Process parameters	Units	Low level (−1)	High level (+1)
A	Flux density	g/c/c	0.85	0.90
B	Conveyor speed	ft./min.	4.5	5.5
C	Solder temperature	°C	230	260

Table 3.2 OVAT approach to wave-soldering process

Run	A	B	C	Response (ppm)
1	−1	−1	−1	420
2	+1	−1	−1	370
3	+1	+1	−1	410
4	+1	+1	+1	350

The difference in the responses between the trials 1 and 2 provides an estimate of the effect of process parameter 'A'. From Table 3.2, the effect of 'A' $(370 − 420 = −50)$ was estimated when the levels of 'B' and 'C' were at low levels. There is no guarantee whatsoever that 'A' will have the same effect for different conditions of 'B' and 'C'. Similarly, the effects of 'B' and 'C' can be estimated. In the above experiment, the response values corresponding to the combinations A (−1) B (+1), A (−1) C (+1) and B (−1) C (+1) are missing. Therefore, OVAT to experimentation can lead to unsatisfactory conclusions and in many cases it would even lead to false optimum conditions. In this case, the team failed to study the effect of each factor at different conditions of other factors. In other words, the team failed to study the interaction between the process parameters.

Interactions occur when the effect of one process parameter depends on the level of the other process parameter. In other words, the effect of one process parameter on the response is different at different levels of the other process parameter. In order to study interaction effects among the process parameters, we need to vary all the factors simultaneously. For the above wave soldering process, the engineering team has employed a Full Factorial Experiment (FFE)

Table 3.3 Results from a 2^3 full factorial experiment

Run (standard order)	Run (randomized order)	A	B	C	Response (ppm)
1	5	−1	−1	−1	420, 412
2	7	+1	−1	−1	370, 375
3	4	−1	+1	−1	310, 289
4	1	+1	+1	−1	410, 415
5	8	−1	−1	+1	375, 388
6	3	+1	−1	+1	450, 442
7	2	−1	+1	+1	325, 322
8	6	+1	+1	+1	350, 340

Table 3.4 Average ppm values

Run (standard order)	A	B	Average ppm
1, 5	− 1	− 1	398.75
3, 7	− 1	+ 1	311.50
2, 6	+ 1	− 1	409.25
4, 8	+ 1	+ 1	378.75

and each trial or run condition was replicated twice to observe variation in results within the experimental trials. The results of the FFE are shown in Table 3.3. Each trial condition was randomized to minimize the effect of undesirable disturbances or external factors which are uncontrollable or expensive to control during the experiment.

As it is a FFE, it is possible to study all the interactions among the factors A, B and C. The interaction between two process parameters (say, A and B) can be computed using the following equation:

$$I_{A,B} = \frac{1}{2}\left(E_{A,B(+1)} - E_{A,B(-1)}\right) \qquad (3.1)$$

where $E_{A,B\ (+1)}$ is the effect of factor 'A' at high level of factor 'B' and where $E_{A,B\ (-1)}$ is the effect of factor 'A' at low level of factor 'B'.

For the above example, three two-order interactions and a third-order interaction can be studied. Third-order and higher order interactions are not often important for process optimization problems and therefore not necessary to be studied. In order to study the interaction between A (flux density) and B (conveyor speed), it is important to form a table (Table 3.4) for average ppm values at the four possible combinations of A and B (i.e. $A_{(-1)}\ B_{(-1)}$, $A_{(-1)}\ B_{(+1)}$, $A_{(+1)}\ B_{(-1)}$ and $A_{(+1)}\ B_{(+1)}$).

From the above table, effect of 'A' (i.e. going from low level (−1) to high level

$$(+1) \text{ at high level of 'B' (i.e. } + 1) = 378.75 - 311.50$$
$$= 67.25\,\text{ppm}$$

Similarly, effect of A at low level of B $= 409.25 - 398.75$
$$= 10.5\,\text{ppm}$$

Interaction between A and B $= \frac{1}{2}[67.25 - 10.5]$
$$= 28.375$$

In order to determine whether two process parameters are interacting or not, one can use a simple but powerful graphical tool called interaction graphs. If the lines in the interaction plot are parallel, there is no interaction between the process parameters. This implies that the change in the mean

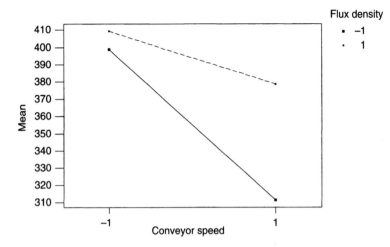

Figure 3.1 Interaction plot between flux density and conveyor speed.

response from low to high level of a factor does not depend on the level of the other factor. On the other hand, if the lines are non-parallel, an interaction exists between the factors. The greater the degree of departure from being parallel, the stronger the interaction effect. Figure 3.1 illustrates the interaction plot between 'A' (flux density) and 'B' (conveyor speed).

The interaction graph between flux density and conveyor speed shows that the effect of conveyor speed on ppm at two different levels of flux density is not the same. This implies that there is an interaction between these two process parameters. The defect rate (in ppm) is minimum when the conveyor speed is at high level and flux density at low level.

3.2 Alternative method for calculating the two order interaction effect

In order to compute the interaction effect between flux density and conveyor speed, we need to first multiply columns 2 and 3 in Table 3.4. This is

Table 3.5 Alternative method to compute the interaction effect

A	B	A × B	Average ppm
−1	−1	+1	398.75
−1	+1	−1	311.50
+1	−1	−1	409.25
+1	+1	+1	378.75

illustrated in Table 3.5. In Table 3.5, column 3 yields the interaction between flux density (A) and conveyor speed (B).

Having obtained column 3, we then need to calculate the average ppm at high level of (A × B) and low level of (A × B). The difference between these will provide an estimate of the interaction effect.

$$A \times B = \text{Average ppm at high level of } (A \times B)$$

$$- \text{Average ppm at low level of } (A \times B)$$

$$= \frac{1}{2}(398.75 + 378.75) - \frac{1}{2}(311.50 + 409.25)$$

$$= 388.75 - 360.375$$

$$= 28.375$$

Now consider the interaction between flux density (A) and solder temperature. The interaction graph is shown in Figure 3.2. The graph shows that the effect of solder temperature at different levels of flux density is almost same. Moreover the lines are almost parallel, which indicates that there is little interaction between these two factors.

The interaction plot suggests that the mean solder defect rate is minimum when solder temperature is at high level and flux density at low level. *Note*: Non-parallel lines is an indicator of the existence of interactions between two factors and parallel lines indicate no interactions between the factors.

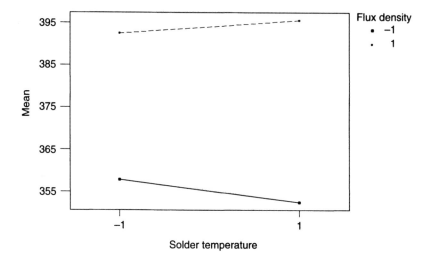

Figure 3.2 Interaction plot between solder temperature and flux density.

3.3 Synergistic interaction vs antagonistic interaction

The effects of process parameters can be either fixed or random. Fixed process parameter effects occur when the process parameter levels included in the experiment are controllable and specifically chosen because they are the only ones for which inferences are desired. For example, if you want to determine the effect of temperature at two-levels (180 °F and 210 °F) on the viscosity of a fluid, then both 180 °F and 210 °F are considered to be fixed parameter levels. On the other hand, random process parameter effects are associated with those parameters whose levels are randomly chosen from a large population of possible levels. Inferences are not usually desired on the specific parameter levels included in an experiment, but on the population of levels represented by those in the experiment. Factor levels represented by batches of raw materials drawn from a large population are examples of random process parameter levels. In this book, only fixed process parameter effects are considered.

For synergistic interaction, the lines on the plot do not cross each other. For example, Figure 3.1 is an example for synergistic interaction. In contrast, for antagonistic interaction, the lines on the plot cross each other. This can be illustrated in Figure 3.3. In this case, the change in mean response for factor B at low level (represented by −1) is noticeably high compared to high level. In other words, factor B is less sensitive to variation in mean response at high level of factor A.

In order to have a greater understanding of the analysis and interpretation of interaction effects, the following two scenarios can be considered.

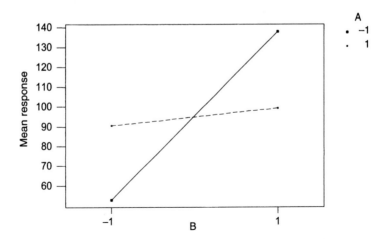

Figure 3.3 Antagonistic interaction between two factors A and B.

3.4 Scenario 1

In an established baking school, the students had failed to produce uniform-sized cakes, despite their continuous efforts. The engineering team of the company was looking for the key factors or interactions which are most responsible for the variation in the weight of cakes. Here the weight of the cakes was considered to be the critical characteristic to the customers. A project was initiated to understand the nature of the problem and come up with a possible solution to identify the causes of variation and if possible eliminate them for greater consistency in the weights of these cakes. After a thorough brainstorming session, six process variables (or factors) and a possible interaction (B × M) were considered for the experiment. The factors and their levels are shown in Table 3.6.

Each process variable was kept at 2-levels and the objective of the experiment was to determine the optimum combination of process variables which yields minimum variation in the weight of cakes. A FFE would have required 64 experimental runs. Due to limited time and experimental budget, it was decided to select a $2^{(6-3)}$ (i.e. eight trials or runs). Each trial condition was replicated twice for obtaining sufficient degrees of freedom for the error term. Because we are analysing variation, the minimum number of replicates per trial condition is two. Table 3.7 shows the experimental layout or design matrix for the cake baking experiment. According to Central Limit Theorem (CLT), if you repeatedly take large random samples from a stable process and display the averages of each sample in a frequency diagram, the diagram will be approximately bell-shaped. In other words, the sampling distribution of means is roughly normal, according to CLT. It is quite interesting to note that the distribution of sample Standard Deviations (SD) does not follow a normal distribution. However, if we transform the sample SD by taking their logarithms, the logarithms of the SD will be much closer to being normally distributed. The last column in Table 3.7 gives the logarithmic transformation of sample SD. The SD and log(SD) can be easily obtained by using a scientific calculator or Microsoft Excel spreadsheet. Here our interest is to analyse the interaction between the process variables butter (B) and milk (M) rather than the individual effect of each process variable on the variability of cake weights.

Table 3.6 List of baking process variables for the experiment

Factors	Label	Low level	High level
Butter (cups)	B	1/4	1/2
Milk (cups)	M	1/4	1/2
Flour (cups)	F	3/4	1
Sugar (cups)	S	1/2	3/4
Oven temperature (°C)	O	200	225
Eggs	E	2	3

Table 3.7 Response table for the cake baking experiment

Run	B	M	B × M	O	F	S	E	Weight (grams)	log(SD)
1	−1	−1	+1	−1	+1	+1	−1	102.3, 117.6	1.034
2	+1	−1	−1	−1	−1	+1	+1	114.6, 120.3	0.605
3	−1	+1	−1	−1	+1	−1	+1	134.6, 126.7	0.747
4	+1	+1	+1	−1	−1	−1	−1	116.4, 123.9	0.725
5	−1	−1	+1	+1	−1	−1	+1	112.6, 130.6	1.105
6	+1	−1	−1	+1	+1	−1	−1	150.6, 141.7	0.799
7	−1	+1	−1	+1	−1	+1	−1	133.6, 122.4	0.899
8	+1	+1	+1	+1	+1	+1	+1	155.8, 138.6	1.085

In order to analyse the interaction effect between butter and milk, we form a table for average log(SD) values corresponding to all the four possible combinations of B and M. The results are shown in Table 3.8.

Calculation of interaction effect (B × M):

$$\text{Effect of butter (B) at high level of milk (M)} = 0.905 - 0.823$$
$$= 0.082$$

$$\text{Effect of butter (B) at low level of milk (M)} = 0.702 - 1.0695$$
$$= -0.3675$$

Using Eq. (4.1),

$$B \times M = \frac{1}{2}[0.082 - (-0.3675)]$$
$$= \frac{1}{2}[0.082 + 0.3675] = 0.225$$

Figure 3.4 illustrates the interaction plot between the process variables 'B' and 'M'.

Figure 3.4 clearly indicates the existence of interaction between the factors butter and milk. The interaction plot shows that variability in the weight of cakes is minimum when the level of butter is kept at high level and milk at low level.

Table 3.8 Interaction table for log(SD)

B	M	Average log(SD)
−1	−1	11.0695
−1	+1	0.823
+1	−1	0.702
+1	+1	0.905

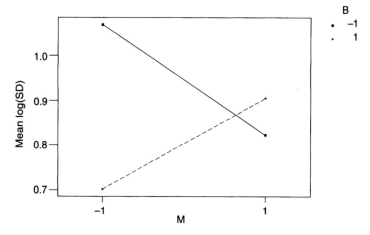

Figure 3.4 Interaction plot between milk and butter.

3.5 Scenario 2

In this scenario, we illustrate an experiment conducted by a chemical engineer to study the effect of three process variables (temperature, catalyst and pH) on the chemical yield. The results of the experiment are shown in Table 3.9. The engineer was interested to study the effect of three process variables and the interaction between temperature and catalyst. The engineer has replicated each trial condition three times for obtaining sufficient degrees of freedom for the experimental error. Moreover, replication increases the precision of the experiment by reducing the standard deviations used to estimate the process parameter (or factor) effects.

The first step was to construct a table (Table 3.10) for interaction between TE and CA. The mean chemical yield at all four combinations of TE and CA was estimated. In order to determine whether or not these variables are interacting, an interaction plot was constructed (Figure 3.5).

Table 3.9 Experimental layout for the yield experiment

Trial	TE	CA	pH	Chemical yield (%)
1	−1	−1	−1	60.4, 62.1, 63.4
2	+1	−1	−1	64.1, 79.4, 74.0
3	−1	+1	−1	59.6, 61.2, 57.5
4	+1	+1	−1	66.7, 67.3, 68.9
5	−1	−1	+1	63.3, 66.0, 65.3
6	+1	−1	+1	91.2, 77.4, 84.9
7	−1	+1	+1	68.1, 71.3, 68.6
8	+1	+1	+1	75.3, 77.1, 76.1

Table 3.10 TE × CA Interaction table

TE	CA	Mean chemical yield
−1	−1	63.42
+1	−1	78.50
−1	+1	64.38
+1	+1	71.90

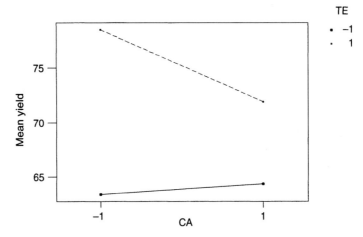

Figure 3.5 Interaction plot between CA and TE.

As the lines are not parallel, there is an interaction between the process variables CA and TE. The graph indicates that the effect of CA is insensitive to mean yield at low level of TE. However, maximum yield is obtained when temperature is kept at high level and CA at low level. The interaction effect can be computed in the following manner.

Effect of CA at high level of

$$TE = 71.90 - 78.50 = -6.60$$

Effect of CA at low level of

$$TE = 64.38 - 63.42 = 0.96$$

$$CA \times TE = \frac{1}{2}[-6.60 - 0.96] = -3.78$$

3.6 Summary

This chapter illustrates the significance of interactions in industrial processes and how to deal with them. In order to study and analyse interactions among the process or design parameters, we have to vary them at their respective levels simultaneously. In order to understand the presence of interaction between two process parameters, it is encouraged to employ a simple and powerful graphical tool called interaction graph or plot. If the lines in the plot are parallel, it implies no interaction between the process parameters. In contrast, non-parallel lines is an indication of the presence of interaction. The chapter also presents two scenarios for better and rapid understanding of how to interpret interactions in industrial experiments.

Exercises

1. In a certain casting process for manufacturing jet engine turbine blades, the objective of the experiment is to determine the most important interaction effects (if there are any) that affect the part shrinkage. The experimenter has selected three process parameters: pour speed (A), metal temperature (B) and mould temperature (C), each factor being kept at two levels for the study. The response table, together with the response values, is shown below. Calculate and analyse the two-factor interactions among the three process variables. Each run was replicated three times to have adequate degrees of freedom for error.

Run	A	B	C	Shrinkage
1	−1	−1	−1	2.22, 2.11, 2.14
2	+1	−1	−1	1.42, 1.54, 1.05
3	−1	+1	−1	2.25, 2.31, 2.21
4	+1	+1	−1	1.00, 1.38, 1.19
5	−1	−1	+1	1.73, 1.86, 1.79
6	+1	−1	+1	2.71, 2.45, 2.46
7	−1	+1	+1	1.84, 1.76, 1.70
8	+1	+1	+1	2.27, 2.69, 2.71

2. A company that manufactures can-forming equipment wants to set up an experiment to help understand the factors influencing surface finish on a particular steel subassembly. The company decides to perform an eight trial experiment with three factors at two levels. A brainstorming session conducted with people within the organization such as operator, supervisor and engineer resulted in the finished part being measured at four places. The list of factors (A: tool radius, B: feed rate and C: RPM) and the response (surface finish) are shown in the following experimental layout. Generate an interaction plot for any two-way interactions with large effects.

Run	A	B	C	Surface finish
1	− 1	− 1	− 1	50, 50, 55, 50
2	+ 1	− 1	− 1	145, 150, 100, 110
3	− 1	+ 1	− 1	160, 165, 155, 160
4	+ 1	+ 1	− 1	180, 200, 190, 195
5	− 1	− 1	+ 1	60, 65, 55, 60
6	+ 1	− 1	+ 1	25, 35, 35, 30
7	− 1	+ 1	+ 1	160, 160, 150, 165
8	+ 1	+ 1	+ 1	80, 70, 75, 80

References

Anderson, M.J. and Whitcomb, P.J. (2000). *DOE Simplified: Practical Tools for Effective Experimentation*. Portland, Oregon, USA, Productivity Inc.

Antony, J. and Kaye, M. (1998). Key Interactions. *Manufacturing Engineer*, June, 77(3), 136–138.

Barton, R. (1999). *Graphical Methods for the Design of Experiments*. NY, USA, Springer-Verlag.

Gunst, R.F. and Mason, R.L. (1991). *How to Construct Fractional Factorial Experiments*. Milwaukee, Wisconsin, USA, ASQC Quality Press.

Lochner, R.H. and Matar, J.E. (1990). *Designing for Quality – An Introduction to the Best of Taguchi and Western Methods of Experimental Design*. London, UK, Chapman and Hall Publishers.

Logothetis, N. (1994). *Managing for Total Quality*. NY, USA, Prentice-Hall.

4

A systematic methodology for Design of Experiments

4.1 Introduction

It is widely considered that DOE or Experimental Design forms an essential part of the quest for effective improvement in process performance or product quality. This chapter discusses the barriers and cognitive gaps in the statistical knowledge required by industrial engineers for tackling process- and quality-related problems using DOE technique. This chapter also presents a systematic methodology to guide people in organizations with limited statistical ability for solving manufacturing process-related problems in real life situations.

4.2 Barriers in the successful application of DOE

The 'effective' application of DOE by industrial engineers is limited in many manufacturing organizations. Some noticeable barriers are:

1. *Educational barriers.* The word 'statistics' invokes fear in many industrial engineers. The fundamental problem begins with the current statistical education for the engineering community in their academic curriculum. The courses currently available in 'engineering statistics' often tend to concentrate on the theory of probability, probability distributions and more mathematical aspects of the subject, rather than practically useful techniques such as DOE, Taguchi method, Robust design, Gauge capability studies, Statistical process control, etc. Engineers must be taught these powerful techniques in the academic world with a number of supporting case studies. This will ensure a better understanding of the application of statistical techniques before they enter the job market.

2. *Management barriers.* Managers often don't understand the importance of DOE in problem solving or don't appreciate the competitive value it brings into the organization. In many organizations, managers encourage their

engineers to use the so called 'home-grown' solutions for process- and quality-related problems. These 'home-grown' solutions are consistent with OVAT approach to experimentation, as managers are always after quick fix solutions which yield short term benefits to their organizations.

3. *Cultural barriers.* Cultural barrier is one of the principal reasons for why DOE is not commonly used in many organizations. The management should be prepared to address all cultural barrier issues that might be present within the organization, plus any fear of training or reluctance to embrace the application of DOE. Many organizations are not culturally ready for the introduction and implementation of advanced quality improvement techniques such as DOE and Taguchi. The best way to overcome this barrier is through intensive training programme and by demonstrating the successful application of such techniques from other organizations during the training.

4. *Communication barriers.* Research has indicated that there is very little communication between the academic and industrial world. Moreover, the communication among industrial engineers, managers and statisticians in many organizations is limited. For the successful initiative of any quality improvement programme, these communities should work together and make this barrier less formidable. For example, lack of statistical knowledge of engineers could lead to problems such as misinterpretation of historical data or misunderstanding of the nature of interactions among factors under consideration for a given experiment. Similarly, academic statisticians' lack of engineering knowledge could lead to problems such as undesirable selection of process variables and quality characteristics for the experiment, lack of measurement system precision and accuracy, etc. Managers' lack of basic knowledge in engineering and statistics could lead to problems such as high quality costs, poor quality and therefore, lost competitiveness in the world market place and so on and so forth.

5. *Other barriers.* Most commercial software systems and expert systems in DOE provide no guidance whatsoever in classifying and analysing manufacturing process quality-related problems from which a suitable approach (Taguchi, classical or Shainin's approach) can be selected. Very little research has been done on this particular aspect and in the author's standpoint, this is probably the most important part of DOE. The selection of a particular approach to experimentation (i.e. Taguchi, classical or Shainin) is dependent upon a number of criteria: complexity involved, degree of optimization required by the experimenter, time required for completion of the experiment, cost issues associated with the experiment, allowed response time to report back to management, etc. Moreover, many software systems in DOE stress data analysis and not properly address data interpretation. Thus, many engineers, having performed the statistical analysis using such software systems, would not know how to utilize the results of the analysis effectively without assistance from statisticians.

4.3 A practical methodology for DOE

The methodology of DOE is fundamentally divided into four phases. These are:

1. planning phase
2. designing phase
3. conducting phase and
4. analysing phase.

4.3.1 Planning phase

The planning phase is made up of the following steps.

(a) *Problem recognition and formulation.* A clear and succinct statement of the problem can create a better understanding of what needs to be done. The statement should contain a specific and measurable objective that can yield practical value to the company. Some manufacturing problems that can be addressed using an experimental approach include:

- Development of new products; improvement of existing processes or products.
- Improvement of the process/product performance relative to the needs and demands of customers.
- Reduction of existing process spread, which leads to poor capability.

Having decided upon the objective(s) of the experiment, an experimentation team can be formed. The team may include a DOE specialist, process engineer, quality engineer, machine operator and a management representative.

(b) *Selection of response or quality characteristic.* The selection of a suitable response for the experiment is critical to the success of any industrialdesigned experiment. The response can be variable or attribute in nature. Variable responses such as length, thickness, diameter, viscosity, strength, etc. generally provide more information than attribute responses such as good/bad, pass/fail or yes/no. Moreover, variable characteristics or responses require fewer samples than attributes require to achieve the same level of statistical significance.

Experimenters should define the measurement system prior to performing the experiment in order to understand what to measure, where to measure, who is doing the measurements, etc. so that various components of variation (measurement system variability, operator variability, part variability, etc.) can be evaluated. It is good to make sure that the measurement system is capable, stable, robust and insensitive to environmental changes.

(c) *Selection of process variables or design parameters.* Some possible ways to identify potential process variables are the use of engineering knowledge of the process, historical data, cause-and-effect analysis and brainstorming. This is a very important step of the experimental design procedure. If important factors are left out of the experiment, then the results of the experiment will not be accurate and useful for any improvement actions. It is good practice to conduct a screening experiment in the first phase of any experimental investigation to identify the most important design parameters or process variables. More information on screening experiments/designs can be obtained from Chapter 5.

(d) *Classification of process variables.* Having identified the process variables, the next step is to classify them into controllable and uncontrollable variables. Controllable variables are those which can be controlled by a process engineer/production engineer in a production environment. Uncontrollable variables (or noise variables) are those which are difficult to control or expensive to control in actual production environments. Variables such as ambient temperature fluctuations, humidity fluctuations, raw material variations, etc. are examples of noise variables. These variables may have some immense impact on the process variability and therefore must be dealt with for enhanced understanding of our process. The effect of such nuisance variables can be minimized by the effective application of DOE principles such as blocking, randomization and replication (For more information on these three principles, refer to Chapter 8: Some useful and practical tips for making your industrial experiments successful.).

(e) *Determining the levels of process variables.* A level is the value that a process variable holds in an experiment. For example, a car's gas mileage is influenced by such levels as tire pressure, speed, etc. The number of levels depends on the nature of the process variable to be studied for the experiment and whether or not the chosen process variable is qualitative (e.g.: type of catalyst, type of material, etc.) or quantitative (temperature, speed, pressure, etc.). For quantitative process variables, two levels are generally required in the early stages of experimentation. However, for qualitative variables, more than two levels may be required. If a non-linear function is expected by the experimenter, then it is advisable to study variables at three or more levels. This would assist in quantifying the non-linear (or curvature) effect of the process variable on the response function.

(f) *List all the interactions of interest.* Interaction among variables is quite common in industrial experiments. In order to effectively interpret the results of the experiment, it is highly desirable to have a good understanding of interaction between two process variables. The best way to relate to interaction is to view as an effect, just like a factor or process variable effect. Since it is not an input you can control, unlike factors or process variables, interactions do not enter into descriptions of trial conditions. In the context of DOE, we generally study two-order interactions. The number of two-order interactions within an experiment can be easily obtained by using a simple equation:

$$N = \frac{n \times (n-1)}{2} \tag{4.1}$$

where n is the number of factors.

For example, if you consider four factors in an experiment, the number of two-order interactions can be equal to six.

The questions to ask include 'do we need to study the interactions in the initial phase of experimentation?', and 'how many two-order interactions are of interest to the experimenter?'. The size of the experiment is dependent on the number of factors to be studied and the number of interactions which are of great concern to the experimenter.

4.3.2 Designing phase

In this phase, one may select the most appropriate design for the experiment. Experiments can be statistically designed using classical approach advocated by Sir Ronald Fisher, orthogonal array approach advocated by Dr Genichi Taguchi or variables search approach promoted by Dr Dorian Shainin. This book is focused on the classical DOE approach advocated by Sir Ronald Fisher. Within this approach, one can choose full factorial, fractional factorial or screening designs (such as Plackett–Burmann designs). These designs are introduced to the reader in the subsequent chapters.

The size of the experiment is dependent on the number of factors and/or interactions to be studied, the number of levels of each factor, budget and resources allocated for carrying out the experiment, etc. During the design stage, it is quite important to consider the confounding structure and resolution of the design. It is good practice to have the design matrix ready for the team prior to executing the experiment. The design matrix generally reveals all the settings of factors at different levels and the order of running a particular experiment.

4.3.3 Conducting phase

This is the phase in which the planned experiment is carried out and the results are evaluated. Several considerations are recognized as being recommended prior to executing an experiment, such as:

- Selection of suitable location for carrying out the experiment. It is important to ensure that the location should not be affected by any external sources of noise (e.g.: vibration, humidity, etc.).
- Availability of materials/parts, operators, machines, etc. required for carrying out the experiment.

- Assessment of the viability of an action in monetary terms by utilising cost–benefit analysis. A simple evaluation must also be carried out in order to verify that the experiment is the only possible solution for the problem at hand and justify that the benefits to be gained from the experiment will exceed the cost of the experiment.

The following steps may be useful while performing the experiment in order to ensure that the experiment is performed according to the prepared experimental design matrix (or layout).

- The person responsible for the experiment should be present throughout the experiment. In order to reduce the operator-to-operator variability, it is best to use the same operator for the entire experiment.
- Monitor the experimental trials. This is to find any discrepancies while running the experiment. It is advisable to stop running the experiment if any discrepancies are found.
- Record the observed response values on the prepared data sheet or directly into the computer.

4.3.4 Analysing phase

Having performed the experiment, the next phase is to analyse and interpret the results so that valid and sound conclusions can be derived. In DOE, the following are the possible objectives to be achieved from this phase:

- Determine the design parameters or process variables that affect the mean process performance.
- Determine the design parameters or process variables that influence performance variability.
- Determine the design parameter levels that yield the optimum performance.
- Determine whether further improvement is possible.

The following tools can be used for the analysis of experimental results. As the focus of this book is to 'Keep It Statistically Simple' for the readers, the author will be introducing only simple but powerful tools for the analysis and interpretation of results. There are a number of DOE books available in the market which cover more sophisticated statistical methods for the analysis. The author encourages readers to use MINITAB software for the analysis of experimental results.

4.4 Analytical tools of DOE

4.4.1 Main effects plot

A main effect plot is a plot of the mean response values at each level of a design parameter or process variable. One can use this plot to compare the relative strength of the effects of various factors. The sign and magnitude of a main effect would tell us the following:

- The sign of a main effect tells us of the direction of the effect, i.e. if the average response value increases or decreases.
- The magnitude tells us of the strength of the effect.

If the effect of a design or process parameter is positive, it implies that the average response is higher at high level than at low level of the parameter setting. In contrast, if the effect is negative, it means that the average response at the low level setting of the parameter is more than at the high level. Figure 4.1 illustrates main effect of temperature on the tensile strength of a steel specimen. As you can see from the figure, tensile strength increases when the setting of temperature varies from low to high (i.e. −1 to 1).

The effect of a process or design parameter (or factor) can be mathematically calculated using the following simple equation:

$$E_f = \bar{F}_{(+1)} - \bar{F}_{(-1)} \qquad (4.2)$$

where $\bar{F}_{(+1)}$ = average response at high level setting of a factor, and $\bar{F}_{(-1)}$ = average response at low level setting of a factor.

4.4.2 Interactions plots

An interactions plot is a powerful graphical tool which plots the mean response of two factors at all possible combinations of their settings. If the lines are parallel, then it connotes that there is an interaction between the factors. Non-parallel lines is an indication of the presence of interaction between the factors. More information on interactions and how to interpret them can be seen in Chapter 3 of the book.

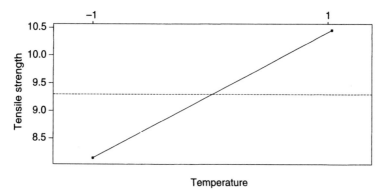

Figure 4.1 Main effect plot of temperature on tensile strength.

4.4.3 Cube plots

Cube plots display the average response values at all combinations of process or design parameter settings. One can easily determine the best and the worst combinations of factor levels for achieving the desired optimum response. A cube plot is useful to determine the path of steepest ascent or descent for optimization problems. Figure 4.2 illustrates an example of a cube plot for a cutting tool life optimization study with three tool parameters; cutting speed, tool geometry and cutting angle. The graph indicates that tool life increases when cutting speed is set at low level and cutting angle and tool geometry are set at high levels. The worst condition occurs when all factors are set at low levels.

4.4.4 Pareto plot of factor effects

The Pareto plot allows one to detect the factor and interaction effects which are most important to the process or design optimization study one has to deal with. It displays the absolute values of the effects, and draws a reference line on the chart. Any effect that extends past this reference line is potentially important. For example, for the above tool life experiment, a Pareto plot is constructed (Figure 4.3). The graph shows that factors B and C and interaction AC are most important. Minitab displays the absolute value of the standardized effects of factors when there is an error term. It is always good practice to check the findings from a Pareto chart with Normal Probability Plot (NPP) of the estimates of the effects (refer to NPP in next section).

4.4.5 Normal Probability Plot of factor effects

For NPPs, the main and interaction effects of factors or process (or design) parameters should be plotted against cumulative probability (per cent). Inactive main and interaction effects tend to fall roughly along a straight line whereas active effects tend to appear as extreme points falling off each end of

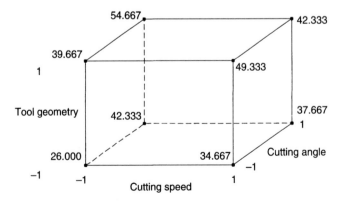

Figure 4.2 Example of a cube plot for cutting tool optimization study.

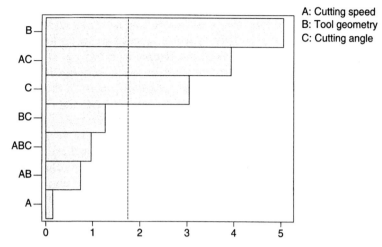

Figure 4.3 Pareto plot of the standardized effects.

the straight line. These active effects are judged to be statistically significant. Figure 4.4 shows a NPP of effects of factors for the above cutting tool optimization example at 5 per cent significance level. Here significance level is the risk of saying that a factor is significant when in fact it is not. In other words, it is the probability of the observed significant effect being due to pure chance. The results are absolutely identical to that of Pareto plot of factor/interaction effects.

4.4.6 Normal Probability Plot of residuals

One of the key assumptions for the statistical analysis of data from industrial experiments is that the data come from a normal distribution. The appearance

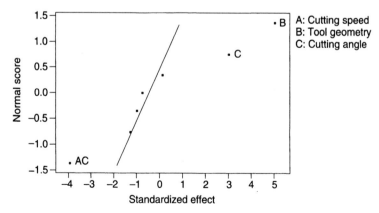

Figure 4.4 Normal Probability Plot of effects for cutting tool optimization example.

of a moderate departure from normality does not necessarily imply a serious violation of the assumptions. Gross deviations from normality are potentially serious and require further analysis. In order to check the data for normality, it is best to construct a NPP of the residuals. Normal probability plots are useful for evaluating the normality of a data set, even when there is a fairly small number of observations. Here a residual means difference in the observed value (obtained from the experiment) and the predicted value or fitted value. If the residuals fall approximately along a straight line, the residuals are then normally distributed. In contrast, if the residuals do not fall fairly close to a straight line, the residuals are then not normally distributed and hence the data do not come from a normal population. The general approach to dealing with non-normality situations is to apply variance-stabilizing transformation on the data. An explanation on data transformation is beyond the scope of this book and therefore readers are advised to refer to Montgomery's book, Design and Analysis of Experiment, which covers the use of data transformation and how to perform data transformation in a detailed manner. Figure 4.5 illustrates the NPP of residuals for the cutting tool optimization example. The graph shows that the points fall fairly close to a straight line indicating that the data are approximately normal.

4.4.7 Response surface plots and regression models

Response surface plots such as contour and surface plots are useful for establishing desirable response values and operating conditions. In a contour plot, the response surface is viewed as a two-dimensional plane where all points that have the same response are connected to produce contour lines of constant responses. A surface plot generally displays a three-dimensional view that may provide a clearer picture of the response. If the regression model (i.e. first-order model) contains only the main effects and no inter-

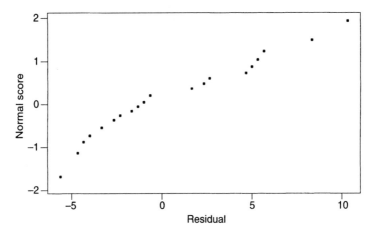

Figure 4.5 Normal probability plot of residuals for the cutting tool example.

action effect, the fitted response surface will be a plane (i.e. contour lines will be straight). If the model contains interaction effects, the contour lines will be curved and not straight. The contours produced by a second-order model will be elliptical in nature. Figures 4.6 and 4.7 illustrate the contour and surface plots of cutting tool life (hours).

Both contour and surface plots help experimenters to understand the nature of the relationship between the two factors (cutting speed and cutting angle) and the response (life in hours). As can be seen in Figures 4.6 and 4.7, the tool life increases with increase in cutting angle and decrease in cutting speed. Moreover, we have used a fitted surface (Figure 4.7) to find a direction of potential improvement for a process. A formal way to seek the direction of improvement in process optimization problems is called the method of steepest ascent or descent (depending on the nature of the problem at hand, i.e. whether one needs to maximize or minimize the response of interest).

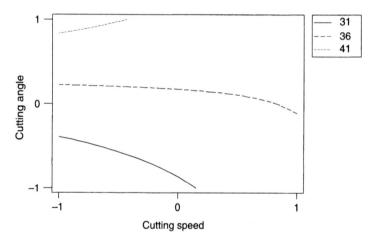

Figure 4.6 Contour plot of cutting tool life.

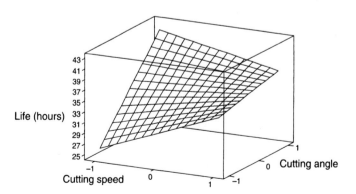

Figure 4.7 Surface plot of cutting tool life.

4.5 Model building for predicting response function

This section is focused on the model building and prediction of response function at various operating conditions of the process. Here the author uses a regression model approach to illustrate the relationship between a response and a set of process parameters (or design parameters) which affect the response. The use of this regression model is to predict the response for different combinations of process parameters (or design parameters) at their best levels. In order to develop a regression model based on the significant effects (either main or interaction), the first step is to determine the regression coefficients. For factors at 2-levels, the regression coefficients are obtained by dividing the estimates of effects by 2. The reason is that a two unit change (i.e. low level setting (−1) to a high level setting (+1)) in a process parameter (or factor) produces a change in the response function. A regression model for factors at 2-levels is usually of the form:

$$\hat{y} = \beta_0 + \beta_1 x_1 + \beta_2 x_2 + \cdots + \beta_{12} x_1 x_2 + \beta_{13} x_1 x_3 + \cdots + \varepsilon \qquad (4.3)$$

where $\beta_1, \beta_2 \ldots$ are the regression coefficients and β_0 is the average response in a factorial experiment. The term 'ε' is the random error component which is approximately normally and independently distributed with mean zero and constant variance σ^2. The regression coefficient β_{12} corresponds to the interaction between the process parameters x_1 and x_2. For example, the regression model for the cutting tool life optimization study is given by:

$$\hat{y} = 40.833 + 5.667(B) + 3.417(C) - 4.417(AC) \qquad (4.4)$$

The response values obtained from Eq. (4.4) are called predicted values and the actual response values obtained from the experiment are called observed values. Residuals can be obtained by taking the difference of observed and predicted (or fitted) values. Equation (4.4) provides us with a tool that can be used to study the response as a function of three tool life parameters; cutting speed, tool geometry and cutting angle. We can predict the cutting tool life for various combinations of these tool parameters. For instance, if all the cutting tool life parameters are kept at low level settings, the predicted tool life then would be:

$$\begin{aligned} \hat{y} &= 40.833 + 5.667(B) + 3.417(C) - 4.417(AC) \\ &= 40.833 + 5.667(-1) + 3.417(-1) - 4.417(-1) \times (-1) \\ &= 27.332 \end{aligned}$$

The observed value of tool life (refer to cube plot) is 26 hours. The difference between the observed value and predicted value (i.e. residual) is −1.332. Similarly, if all the cutting tool life parameters are kept at the optimal condition (i.e. cutting speed = low, tool geometry = high and cutting angle = high), the predicted tool life would then be:

$$\hat{y} = 40.883 + 5.667(+1) + 3.417(+1) - \{4.417(-1) \times (+1)\}$$
$$= 54.384$$

Once the statistical analysis is performed on the experimental data, it is important to verify the results by means of confirmatory experiments or trials. The number of confirmatory runs at the optimal settings can vary from 4 to 20 (4 runs if expensive, 20 runs if cheap).

4.6 Confidence interval for the mean response

The statistical confidence interval (at 99 per cent confidence limit) for the mean response can be computed using the equation:

$$CI = \bar{y} \pm 3 \left\{ \frac{s}{\sqrt{n}} \right\} \tag{4.5}$$

where $\bar{y}$ = mean response obtained from confirmation trials or runs, SD = standard deviation of response obtained from confirmation trials, and n = number of samples (or confirmation runs).

For the cutting tool life example, five samples were collected from the process at the optimal condition (i.e. cutting speed = low, tool geometry = high and cutting angle = high). The results of the confirmation trials are illustrated in Table 4.1.

$$\bar{y} = 53.71 \text{ hours and SD} = 0.654 \text{ hours}$$

Ninety-nine per cent confidence interval for the mean response is given by:

$$CI = 53.71 \pm 3 \left\{ \frac{0.654}{\sqrt{5}} \right\}$$
$$= 53.71 \pm 0.877 = (54.55, 52.83)$$

As the predicted value based on the regression model falls within the statistical confidence interval, we will consider our model good!

If the results from the confirmation trials or runs fall outside the statistical confidence interval, possible causes must be identified. Some of the possible causes may be:

- incorrect choice of experimental design for the problem at hand
- improper choice of response(s) for the experiment
- inadequate control of noise factors, which causes excessive variation

Table 4.1 Confirmation trials

Results from confirmation trials
53.48
52.69
53.88
54.12
54.36

- some important process or design parameters which have been omitted in the first rounds of experimentation
- measurement error
- wrong assumptions regarding interactions
- errors in conducting the experiment, etc.

If the results from the confirmatory trials or runs are within the confidence interval, then improvement action on the process is recommended. The new process or design parameters should be implemented with the involvement of top management. After the solution has been implemented, control charts on the response(s) or key process parameters should be constructed for constantly monitoring, analysing, managing and improving the process performance.

4.7 Summary

Industrially designed experiments do not always go as planned because a non-systematic approach is often taken by the experimenters and scientists in organizations. The purpose of this chapter is to provide the necessary steps for planning, designing, conducting and analysing industrially designed experiments in a disciplined and structured manner. The chapter also presents the common barriers in the successful implementation of DOE in many organizations.

Exercises

1. What are the common barriers in the successful application of DOE?
2. Discuss the four phases in the methodology of DOE
3. What are the criteria for the selection of an experimental design?
4. Explain the key considerations which need to be taken into account prior to executing an experiment
5. What is the use of NPP of residuals?
6. Explain the role of response surface plots in industrial experiments
7. Why do we need to develop regression models?
8. What are the possible causes of experiments being unsuccessful?

References

Antony, J. and Kaye, M. (1995). A Methodology for Taguchi Design of Experiments for continuous quality improvement. *Quality World Technical Supplement*, September, pp. 98–102.

Benski, H.C. (1989). Use of a normality test to identify significant effects in factorial designs. *Journal of Quality Technology*, 21(3), 174–178.

Kumar, S. and Tobin, M. (1990). Design of Experiments is the best way to optimise a process at minimal cost. *IEEE/CHMT*, pp. 166–173.

Launsby, R. and Weese, D. (1995). *Straight Talk on Designing Experiments*. Colorado Springs, Colorado, USA, Launsby Consulting.

Marilyn, H. (1993). A Holistic approach to the design of experiments, *ASQC, Statistics Division Newsletter*, 3(13), 16–20.

Minitab Statistical Software User Manual (2000). Release 13 for Windows. (February).

Montgomery, D.C. (2001). *Design and Analysis of Experiments*. USA, John Wiley and Sons.

5

Screening designs

5.1 Introduction

In many process development and manufacturing applications, the number of potential process or design (factors) is large. Screening is used to reduce the number of process or design parameters (or factors) by identifying the key ones that affect product quality or process performance. This reduction allows one to focus process improvement efforts on the few really important factors, or the 'vital few'.

Screening designs provide an effective way to consider a large number of process or design parameters (or factors) in a minimum number of experimental runs or trials (i.e. minimum resources and budget). The purpose of screening designs is to identify and separate out those factors that demand further investigation. This chapter is focused on the screening designs expounded by R.L. Plackett and J.P. Burman in 1946 and hence the name Plackett–Burman designs (P–B designs). The P–B designs are based on Hadamard matrices in which the number of experimental runs or trials is a multiple of four, i.e. $N = 4$, 8, 12, 16, ... and so on, where N is the number of trials/runs.

P–B designs are suitable for studying up to $k = (N-1)/(L-1)$ factors, where L is the number of levels and k is the number of factors. For instance, using a 12 run experiment, it is possible to study up to 11 process or design parameters at 2-levels. One of the interesting properties of P–B designs is that all main effects are estimated with the same precision. This implies that one does not have to anticipate which factors are most likely to be important when setting up the study. For screening designs, experimenters are generally not interested to investigate the nature of interactions among the factors. The aim is to study as many factors as possible in a minimum number of trials and identifying those that need to be studied in further rounds of experimentation in which interactions can be more thoroughly assessed.

5.2 Geometric and non-geometric P–B designs

Geometric P–B designs are those in which N is a power of two. The number of runs can be 4, 8, 16, 32, etc. Geometric designs are identical to fractional factorial designs (refer to Chapter 7) in which one may be able to study the interactions between factors. For example, an 8 run geometric P–B design is shown in Table 5.1. This allows one to study up to 7 factors at 2-levels.

Each P–B design can be easily constructed using a 'generating vector' which, for example, in the case of $N=4$ has the form $(-1 +1 +1)$. The design matrix or experimental layout is obtained by arranging the vector as the first column and off-setting by one vector element for each new column. In other words, a new column is generated from the previous one by moving the elements of the previous column down once, and placing the last element in the first position. The matrix is completed by a row of ones. Table 5.2 illustrates the competed design matrix for a four run P–B design ($N=4$) using the above generating vector.

Non-geometric P–B designs are designs which are multiples of four but are not powers of two. Such designs have runs of 12, 20, 24, 28, etc. These designs do not have complete confounding of effects. For non-geometric P–B designs, each main effect is partially confounded with all interactions that do not contain the main effect. If the interaction effect is suspected to be large, then the interaction may distort the estimated effects of several process or design parameters, since each interaction is partially confounded with all main effects except the two interacting factors. Table 5.3 illustrates the design matrix for a 12 run non-geometric P–B design with generating vector: $(+1 +1 -1 +1 +1 +1 -1 -1 -1 +1 -1)$. This design should not be used to analyse interactions. A 12 run P–B design is generally used for studying 11 main effects. There is nothing wrong having fewer than 11 factors. If the process is suspected to be highly interactive, it would be better to use

Table 5.1 An 8 run geometric P–B design

A	B	C	D	E	F	G
+1	−1	−1	+1	−1	+1	+1
+1	+1	−1	−1	+1	−1	+1
+1	+1	+1	−1	−1	+1	−1
−1	+1	+1	+1	−1	−1	+1
+1	−1	+1	+1	+1	−1	−1
−1	+1	−1	+1	+1	+1	−1
−1	−1	+1	−1	+1	+1	+1
−1	−1	−1	−1	−1	−1	−1

Table 5.2 Design matrix for a 4 run geometric P–B design

A	B	C
−1	+1	+1
+1	−1	+1
+1	+1	−1
−1	−1	−1

Table 5.3 A 12 run non-geometric P–B design

A	B	C	D	E	F	G	H	I	J	K
+1	−1	+1	−1	−1	−1	+1	+1	+1	−1	+1
+1	+1	−1	+1	−1	−1	−1	+1	+1	+1	−1
−1	+1	+1	−1	+1	−1	−1	−1	+1	+1	+1
+1	−1	+1	+1	−1	+1	−1	−1	−1	+1	+1
+1	+1	−1	+1	+1	−1	+1	−1	−1	−1	+1
+1	+1	+1	−1	+1	+1	−1	+1	−1	−1	−1
−1	+1	+1	+1	−1	+1	+1	−1	+1	−1	−1
−1	−1	+1	+1	+1	−1	+1	+1	−1	+1	−1
−1	−1	−1	+1	+1	+1	−1	+1	+1	−1	+1
+1	−1	−1	−1	+1	+1	+1	−1	+1	+1	−1
−1	+1	−1	−1	−1	+1	+1	+1	−1	+1	+1
−1	−1	−1	−1	−1	−1	−1	−1	−1	−1	−1

a geometric design as opposed to a non-geometric design. In contrast, if interactions are of no concern to the experimenter, it is advisable to use a non-geometric design.

The generating vectors for P–B designs are as follows:

$$N = 4 \; (-1 + 1 + 1)$$
$$N = 8 \; (+1 + 1 + 1 - 1 + 1 - 1 - 1)$$
$$N = 12 \; (+1 + 1 - 1 + 1 + 1 + 1 - 1 - 1 - 1 + 1 - 1)$$
$$N = 16 \; (+1 + 1 + 1 + 1 - 1 + 1 - 1 + 1 + 1 - 1 - 1 + 1 - 1 - 1 - 1)$$
$$N = 20 \; (+1 + 1 - 1 - 1 + 1 + 1 + 1 + 1 - 1 + 1 - 1 + 1 - 1 - 1 - 1$$
$$- 1 + 1 + 1 - 1)$$

The obvious advantage of P–B designs is the limited number of runs to evaluate large number of factors. Since interactions are not of interest to the experimenter for P–B designs, the important main effects can be selected for more in-depth study. The obvious disadvantage of P–B designs is tied to the assumption required to evaluate up to $k = (N − 1)$ factors in N runs. It is important to note that one can study fewer than $(N − 1)$ factors in N runs. The unused columns can be used to estimate experimental error. Geometric P–B designs are resolution III designs and therefore these designs can be folded over to achieve a design resolution IV.

Example 1
In this section, I would like to illustrate a simple example with an 8 run P–B design which has been used for studying seven factors. The data for this example is taken from Barrentine's book 'An introduction to Design of Experiments: A Simplified Approach'. This example is based on the manufacturing process of a paperboard product. The objective of the experiment was to increase the puncture resistance of this paperboard product. The response or quality characteristic of interest to the team conducting

the experiment was the force required to penetrate the material. The objective is to maximise the mean force required to penetrate the material. Seven factors at two-levels were studied using an 8 run geometric P–B design. Table 5.4 presents the factors selected from the brainstorming session and their levels.

Table 5.4 List of factors and their levels for the experiment

Factors	Labels	Low level setting	High level setting
Paste temperature	A	130 °F	160 °F
Amount of additive	B	0.2%	0.5%
Press roll pressure	C	40 psi	80 psi
Paper moisture	D	Low	High
Paste type	E	No clay	With clay
Cure time	F	10 days	5 days
Machine speed	G	120 fpm	200 fpm

Table 5.5 presents the results of an 8 run geometric P–B design experiment with two replicates per experimental trial condition.

Table 5.5 Design matrix of an 8 run geometric P–B design for the experiment

A	B	C	D	E	F	G	R1	R2
+1	−1	−1	+1	−1	+1	+1	12.5	16.84
+1	+1	−1	−1	+1	−1	+1	42.44	39.29
+1	+1	+1	−1	−1	+1	−1	55.08	47.57
−1	+1	+1	+1	−1	−1	+1	49.37	47.69
+1	−1	+1	+1	+1	−1	−1	55.43	52.80
−1	+1	−1	+1	+1	+1	−1	42.51	35.02
−1	−1	+1	−1	+1	+1	+1	51.13	57.92
−1	−1	−1	−1	−1	−1	−1	15.61	13.65

The data is analysed using Minitab software and the results are illustrated below. The first task is to identify the key main effects which are most influential on the response (i.e. force). Figure 5.1 presents a standardised normal plot of effects for the above experiment. Effects C, E and B fall away from the straight line which implies that they are statistically significant at 5 per cent significance level. Effects A, D, F and G fall along the straight line and therefore can be treated as inactive effects. It is important to note that one can consider even 10 per cent significance level for screening designs in order to ensure that no important factor effects or parameters have been omitted in the first round of experimentation.

In order to substantiate the findings of normal plot, I have used Pareto plot of effects. The Pareto plot (Figure 5.2) shows that effects C (press roll pressure), E (paste type) and B (amount of additive) are most important to

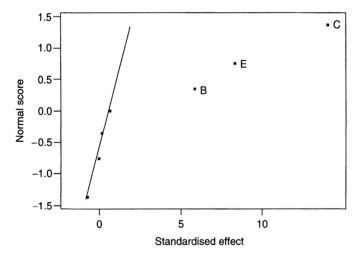

Figure 5.1 Normal probability plot of standardized effects.

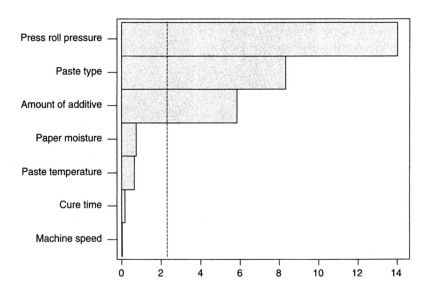

Figure 5.2 Pareto plot of the effects for the experiment.

the process and therefore should be studied at a greater depth. The effect plot of the significant effects is shown in Figure 5.3.

From the above results, one may conclude that main effects C (press roll pressure), E (paste type) and B (amount of additive) are found to have significant impact on the mean puncture resistance (i.e. force required to penetrate the paper board).

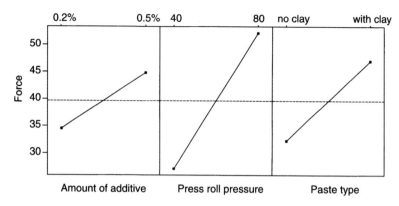

Figure 5.3 Main effects plot of the significant effects.

In order to analyse factors affecting variability in force, we need to calculate SD of observations at each experimental design point. The results are shown in Table 5.6. As we have seen before in the cake baking example (refer to Chapter 3: Understanding key Interactions in processes), the SD of observations do not follow a normal distribution. Therefore, we transform the sample SD by taking their logarithms, as the logarithms of the SD will be much closer to being normally distributed (refer to Chapter 3: Understanding key Interactions in processes). It is important to note that SD can be computed using any scientific calculator.

Figure 5.4 shows a standardized normal plot of effects affecting ln(SD). The normal plot indicates that only factor F (cure time) influenced the variation in the puncture resistance (i.e. force). Further analysis of factor F has revealed that variability is maximum when cure time is set at high level (i.e. 5 days). This can be seen in Figure 5.5.

The conclusions are that factors C, B and E have a significant impact on process average whereas factor F has a significant impact on process

Table 5.6 Design matrix of an 8 run geometric P–B design for the experiment

A	B	C	D	E	F	G	SD	ln(SD)
+1	−1	−1	+1	−1	+1	+1	3.07	1.122
+1	+1	−1	−1	+1	−1	+1	2.23	0.802
+1	+1	+1	−1	−1	+1	−1	5.31	1.670
−1	+1	+1	+1	−1	−1	+1	1.18	0.166
+1	−1	+1	+1	+1	−1	−1	1.86	0.621
−1	+1	−1	+1	+1	+1	−1	5.30	1.668
−1	−1	+1	−1	+1	+1	+1	4.80	1.569
−1	−1	−1	−1	−1	−1	−1	1.39	0.329

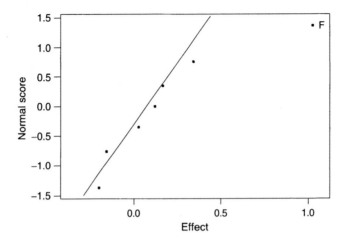

Figure 5.4 Normal plot of effects affecting variability in puncture resistance.

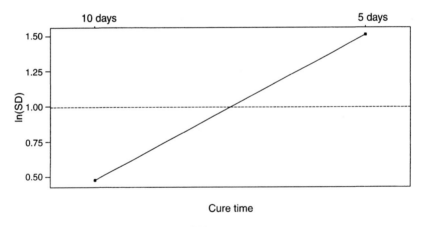

Figure 5.5 Main effects plot for ln(SD).

variability. The other factors such as A, D and G can be set at their economic levels since they do not appear to influence either the process average or the process variability. The next stage of the experimentation would be to consider the interaction among the factors and select the optimal settings from the experiment which yields maximum force with minimum variability. This can be accomplished by utilizing more powerful designs such as full factorials or fractional factorial designs with resolution IV (i.e. main effects are free of third-order interactions or two-factor interactions are confounded with other two-factor interactions).

Example 2

In this example, we consider a plastic foam extrusion process. A process improvement team was formed to investigate what effects porosity of plastic parts. After a thorough brainstorming session with quality engineers, process manager and operators, it was identified that eight process parameters might have some impact on porosity. Table 5.7 presents the list of parameters and their levels for the experiment. Each factor was studied at 2-levels. As the total degrees of freedom for studying 8 factors at 2-levels is equal to 8, it was decided to choose a non-geometric 12 run P–B design with 11 degrees of freedom. The extra 3 degrees of freedom can be used to estimate experimental error. Table 5.8 shows the experimental layout with response values in both standard and random order.

Table 5.7 List of process parameters and their levels for the experiment

Process parameters	Labels	Low level (−1)	High level (+1)
Temperature profile	A	1	2
Temperature after heating	B	210 °C	170 °C
Temperature after expansion	C	170 °C	150 °C
Temperature before coating die	D	130 °C	115 °C
Extrusion speed	E	6 m/min	4.5 m/min
Adhesive coating thickness	F	0.7 mm	0.4 mm
Adhesive coating temperature	G	115 °C	100 °C
Expansion angle	H	Max	Min

Table 5.8 Experimental Layout for 12 run P–B design with response values

Run	A	B	C	D	E	F	G	H	Porosity (%)
1 (6)	+1	+1	−1	+1	+1	+1	−1	−1	44.8
2 (11)	+1	−1	+1	+1	+1	−1	−1	−1	37.2
3 (9)	−1	+1	+1	+1	−1	−1	−1	+1	36.0
4 (7)	+1	+1	+1	−1	−1	−1	+1	−1	34.8
5 (2)	+1	+1	−1	−1	−1	+1	−1	+1	46.4
6 (1)	+1	−1	−1	−1	+1	−1	+1	+1	24.8
7 (5)	−1	−1	−1	+1	−1	+1	+1	−1	43.6
8 (12)	−1	−1	+1	−1	+1	+1	−1	+1	44.8
9 (3)	−1	+1	−1	+1	+1	−1	+1	+1	24.0
10 (8)	+1	−1	+1	+1	−1	+1	+1	+1	34.4
11 (4)	−1	+1	+1	−1	+1	+1	+1	−1	27.2
12 (10)	−1	−1	−1	−1	−1	−1	−1	−1	49.6

Note: Numbers in parentheses represent the random order of experimental runs or trials.

The objective of the experiment was to determine the key parameters which affect percentage porosity. Minitab software system is used for analysis purposes. Figure 5.6 illustrates a standardized Pareto plot of effects for the experiment.

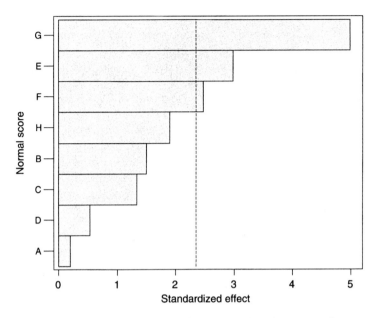

Figure 5.6 Standardized Pareto plot of effects for the above experiment.

Figure 5.6 shows that process parameters such as G (adhesive coating temperature), E (extrusion speed) and F (adhesive coating thickness) have significant impact on porosity. These parameters should be further explored using full fractional designs and more advanced methods such as response surface methods, if necessary. In the next stage of experimentation, one should analyse the interactions among the parameters E, F and G. In order to identify what levels of these parameters yields minimum porosity, we may consider an effects plot (Figure 5.7). Figure 5.7 shows that E at high level, F at low level and G at high level yields minimum porosity.

The figure shows that porosity decreases as temperature is kept at high level (100 °C). Similarly, porosity decreases as extrusion speed is kept at high level (4.5 m/min) and coating thickness at low level (0.7 mm).

5.3 Summary

Screening designs are used for screening a large number of process or design parameters to identify the most important parameters which will have significant impact on the process performance. Once the key parameters are identified, subsequent experimentation can be performed using these parameters to understand and analyse the nature of interactions among them using full/fractional factorial designs and response surface methods, if necessary. Plackett–Burman designs allow the experimenters to evaluate a large

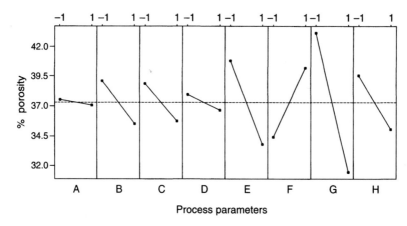

Figure 5.7 Main Effects plot for the experiment.

number of process/design parameters in a minimum number of trials (i.e. minimum budget and resources). One of the stringent assumptions experimenters make is the unimportance of interactions in the early stages of experimentation.

Exercises

1. Compare geometric and non-geometric P–B designs.
2. What are the strengths and limitations of P–B designs?
3. When do you utilize screening designs in real life situations?
4. Explain how do you overcome the problems of low resolution in a screening design?

References

Antony, J. (2002). Training for Design of Experiments using a Catapult. *Quality and Reliability Engineering International*, 18, 29–35.

Barrentine, L.B. (1999). *An introduction to Design of Experiments – A Simplified approach*. Milwaukee, USA, ASQ Quality Press.

Plackett, R.L. and Burmann, J.P. (1946). Design of Optimal Multifactorial Experiments. *Biometrika*, 33, 305–325.

Wheeler, D.J. (1988). *Understanding Industrial Experimentation*. Tennessee, USA, Statistical Process Controls, Inc.

6

Full factorial designs

6.1 Introduction

It is widely accepted that the most commonly used experimental designs in manufacturing companies are full and fractional factorial designs at two-levels and three-levels. Factorial designs would enable an experimenter to study the joint effect of the factors (or process/design parameters) on a response. A factorial design can be either full or fractional factorial. This chapter is primarily focused on full factorial designs at 2-levels only. Factors at 3-levels are beyond the scope of this book. However, if readers wish to learn about experimental design for factors at 3-levels, I would suggest them to refer to Montgomery's book 'Design and Analysis of Experiments'.

A full factorial designed experiment consists of all possible combinations of levels for all factors. The total number of experiments for studying k factors at 2-levels is 2^k. The 2^k full factorial design is particularly useful in the early stages of experimental work, especially when the number of process parameters or design parameters (or factors) is less than or equal to 4. One of the assumptions we make for factors at 2-levels is that the response is approximately linear over the range of the factor settings chosen. The first design in the 2^k series is one with only two factors, say, A and B, each factor to be studied at 2-levels. This is called a 2^2 full factorial design.

6.2 Example of a 2^2 full factorial design

Here we consider a simple nickel plating process with two plating process parameters; plating time and plating solution temperature (refer to Basic Statistics: Tools for Continuous Improvement). Each process parameter is studied at 2-levels. The response of interest to the experimenters was plating thickness. Table 6.1 illustrates the two process parameters and their chosen levels for the experiment.

Table 6.2 shows the design layout of the experiment with response values. Each experimental condition was replicated five times so that a reasonable estimate of error variance (or experimental error) can be obtained.

Table 6.1 Process parameters and their levels for the experiment

Process parameters	Labels	Low level	High level
Plating time	A	4 sec	12 sec
Plating solution temperature	B	16 °C	32 °C

Table 6.2 Design layout of the experiment with response values

Trial number	A	B	Plating thickness				
1	4	16	116.1	116.9	112.6	118.7	114.9
2	4	32	106.7	107.5	105.9	107.1	106.5
3	12	16	116.5	115.5	119.2	114.7	118.3
4	12	32	123.2	125.1	124.5	124.0	124.7

The following are the four objectives set by the experimenter:

1. Which main effects or interactions might affect the mean plating thickness?
2. Which main effects or interactions might influence variability in plating thickness?
3. What is the best setting of factors to minimize variability in thickness?
4. How to achieve a target plating thickness of 120 units?

6.2.1 Objective 1: Determination of main/interaction effects which influence mean plating thickness

In order to determine the effect of process parameters A and B and its interaction AB, we need to construct a coded design matrix with mean plating thickness values as shown in Table 6.3.

The column AB is obtained by simply multiplying the coded values in columns 'A' and 'B'. Interaction AB yields a combined effect of two factors,

Table 6.3 Coded design matrix with mean plating thickness values

A	B	AB	Mean plating thickness
−1	−1	1	115.84
−1	1	−1	106.74
1	−1	−1	116.84
1	1	1	124.30

A and B. The results from Minitab software are shown in Figure 6.1. It illustrates the normal plot of effects. The graph illustrates that process parameter 'plating time' and the interaction between 'plating time and plating solution temperature' are statistically significant at 5 per cent significance level. In other words, these effects have large impact on the mean plating thickness, though plating solution temperature has very little impact on the mean plating thickness. This finding can be further supported by considering the main effects plot and interaction plot (Figures 6.2 and 6.3 respectively).

It can be seen from Figure 6.2 that plating time has a huge impact on plating thickness whereas plating solution temperature has no impact on plating thickness whatsoever. However, it is interesting to note that plating solution temperature has lower sensitivity to variability in plating thickness

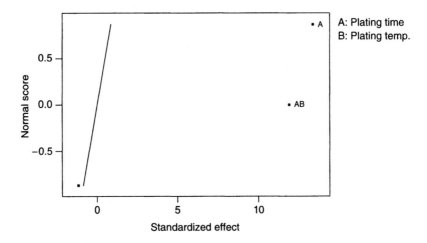

Figure 6.1 Normal probability plot of effects for the plating experiment.

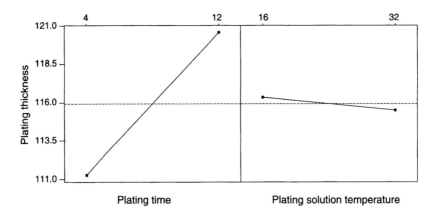

Figure 6.2 Main effects plot for the plating experiment.

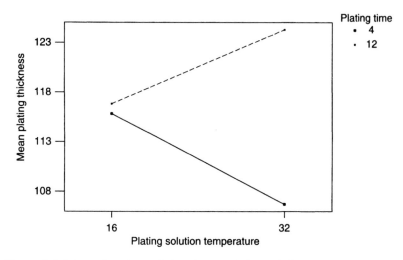

Figure 6.3 Interaction plot – plating time × plating solution temperature.

when compared to plating time. Figure 6.3 indicates that there is a strong interaction between plating time and plating thickness. Plating thickness is maximum when plating time is kept at high level (12 sec) and plating solution temperature is kept at high level (32 °C). Similarly, plating thickness is minimum, when plating solution temperature is kept at high level (32 °C) and plating time is kept at low level (4 sec).

6.2.2 Objective 2: Determination of main/interaction effects which influence variability in plating thickness

In order to determine the effect of A, B and interaction AB on process variability, we need to construct a coded design matrix with response as variability in plating thickness (Table 6.4).

Minitab software is used to identify effects which are most important to process variability. Figure 6.4 shows a Pareto plot of the effects on variability [ln(SD)]. It is quite clear from the graph that process parameter plating solution temperature (B) has a significant effect on plating thickness variability, whereas plating time (A) has no impact on plating thickness variability.

Table 6.4 Coded design matrix with variability as response

A	B	AB	Variability in plating thickness (SD)	ln(SD)
−1	−1	1	2.278	0.823
−1	1	−1	0.607	−0.499
1	−1	−1	1.884	0.633
1	1	1	0.731	−0.313

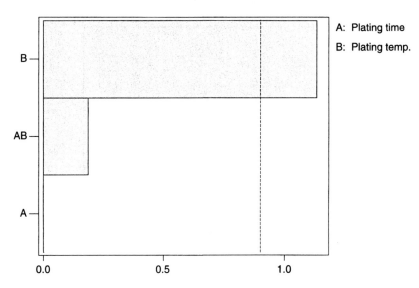

Figure 6.4 Pareto plot of effects on plating thickness variability.

Interaction AB has again very little impact on variability. Figure 6.5 shows that variability is minimum when plating solution temperature is set at high level (32 °C). This finding provides the answer to objective 3 set out earlier in this chapter.

6.2.3 Objective 4: How to achieve a target plating thickness of 120 units?

In order to achieve a target plating thickness of 120 units, we need to initially develop a simple regression model (or mathematical model) which connects the response of interest (i.e. plating thickness) and the significant process

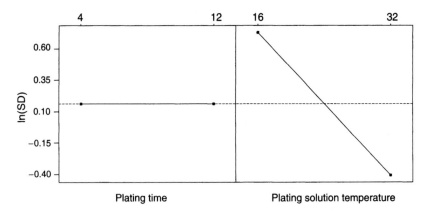

Figure 6.5 Main effects plot with variability as response.

parameters. In order to develop a regression model, we need to construct a table of effects and regression coefficients. It is important to recall that regression coefficients for factors at 2-levels are just half the estimate of effect. A sample calculation of how to estimate the effect of plating time and the interaction between time and temperature is shown below (see Table 6.3).

Effect of plating time on plating thickness

$$\text{Mean plating thickness at high level of plating time} = (116.84 + 124.30)/2$$
$$= 120.57$$
$$\text{Mean plating thickness at low level of plating time} = (115.84 + 106.74)/2$$
$$= 111.29$$
$$\text{Effect of plating time on plating thickness} = (120.57 - 111.29)$$
$$= 9.28$$
$$\text{Regression coefficient of plating time(A)} = 9.28/2$$
$$= 4.64$$

Interaction effect between plating time and plating solution temperature (AB)
Refer to column 3 in Table 7.3,

$$\text{The mean plating thickness at low level of AB} = (106.74 + 116.84)/2$$
$$= 111.79$$

Similarly,

$$\text{The mean plating thickness at high level of AB} = (115.84 + 124.30)/2$$
$$= 120.07$$

Therefore,

$$\text{Interaction AB} = 120.07 - 111.79$$
$$= 8.28$$

$$\text{Regression coefficient of the interaction term AB} = 4.14$$

The regression model for the plating thickness can be therefore written as:

$$\hat{y} = \beta_0 + \beta_1 (A) + \beta_{12} (AB) \tag{6.1}$$

where β_0 = overall mean plating thickness = 115.93, β_1 = regression coefficient of factor A (plating time) and β_{12} = regression coefficient of interaction AB (plating time × plating solution temperature).
The predicted model for plating thickness is therefore given by:

$$\hat{y} = 115.93 + 4.64 (A) + 4.14 (AB)$$

Using the above predicted model, we need to determine the settings of parameters which gives a target thickness of 120 units (i.e. $\hat{y} = 120$). Moreover, we know that high level of plating solution temperature (factor B) yields minimum variability. Therefore, we can set B at low level (i.e. 1).

Now, we can write,

$$120 = 115.93 + 4.64 \ (A) + 4.14 \ (A)$$
$$= 115.93 + 4.46 \ (A) + 4.14 \ (A)$$
$$= 115.93 + 8.78 \ (A)$$
$$4.07 = 8.78 \ (A)$$
$$A = 0.463 \ (\text{in codes terms})$$

The following equation can be used to convert the coded values into actual parameter values (or vice versa).

$$\text{Actual} = \left[\frac{\text{High} + \text{Low}}{2}\right] + \left[\frac{\text{High} - \text{Low}}{2}\right] \cdots \text{Coded} \qquad (6.2)$$

For example, for factor A, high level setting $= 12$ sec, low level setting $= 4$ sec, coded value $= 0.463$

$$\text{Actual} = \{(12 + 4)/2\} + \{((12 - 4)/2)) \ 0.463\}$$
$$= 8 + 4(0.463)$$
$$= 9.85 \text{ sec}$$

Therefore to achieve a target plate thickness of 120 units, we need to set the plating time for 9.85 sec at a temperature of 32 °C. We need to perform confirmation experiments or runs to verify the results of our analysis. If the results of the confirmation experiments or runs (i.e. each observation from the trials) fall within the interval of $\hat{y} \pm 3$ (s.e.), then the results are satisfactory. Here s.e. refers to standard error and is obtained by $SD/\sqrt{n}$, where 'SD' is the sample standard deviation and 'n' sample size.

The analysis of a 2^k factorial design assumes that the observations are normally and independently distributed. The best way to check the normality assumption is by constructing a NPP of residuals. Figure 6.6 presents the normal probability of residuals for the plating experiment. As the residuals fall approximately along a straight line, we can conclude that the data come from a normal population.

6.3 Example of a 2^3 full factorial design

Now we consider an experiment with three factors at 2-levels. The response of interest for the experiment was yield of a chemical process. The list of process parameters and their levels are presented in Table 6.5.

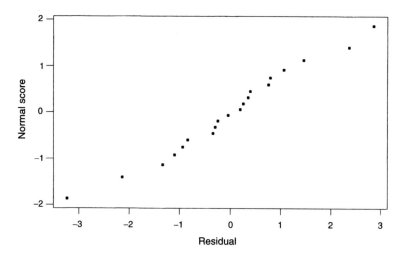

Figure 6.6 Normal probability plot of residuals for the plating experiment.

Table 6.5 List of process parameters and their levels

Process parameters	Labels	Low level	High level
Temperature	T	80 °C	120 °C
Pressure	P	50 psi	70 psi
Reaction time	R	5 min	15 min

It was important to analyse all the two-factor interactions and therefore a 2^3 full factorial design was chosen. Each trial condition was replicated three times in order to obtain an accurate estimate of experimental error (or error variance). The following objectives were set prior to performing the experiment.

1. Which main effects or interactions might affect the average process yield?
2. Which main effects or interactions might influence variability in process yield?
3. What is the optimal process condition?

6.3.1 Objective 1: To identify the significant main/interaction effects which affect the process yield

In order to identify the significant main/interaction effects, it was decided to construct an experimental layout (Table 6.6), which shows all the combinations of process parameters at their respective levels. The table shows the actual settings of the process parameters with the response values (i.e. yield) recorded at each trial condition.

Table 6.6 Experimental layout with response values

Run/trial	T	P	R	Yield 1 (%)	Yield 2 (%)	Yield 3 (%)
1	80	50	5	61.43	58.58	57.07
2	120	50	5	75.62	77.57	75.75
3	80	70	5	27.51	34.03	25.07
4	120	70	5	51.37	48.49	54.37
5	80	50	15	24.80	20.69	15.41
6	120	50	15	43.58	44.31	36.99
7	80	70	15	45.20	49.53	50.29
8	120	70	15	70.51	74.00	74.68

Figure 6.7 illustrates the Pareto plot of effects. The graph shows that main effects T (temperature) and R (reaction time), and interaction between pressure (P) and reaction time (R) are significant at 5 per cent significance level. It is quite interesting to note that pressure (P) on its own has no significant impact on the process yield. It is important to analyse the interaction between P and R for determining the best settings for optimizing the chemical process yield.

Figure 6.8 indicates that there exists a strong interaction between pressure and reaction time. It is clear that the effect of reaction time at different levels of pressure are different. Yield is minimum when the pressure is kept at low

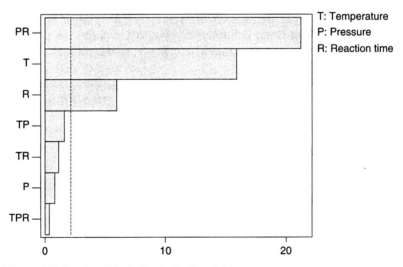

Figure 6.7 Pareto plot of effects for the yield example.

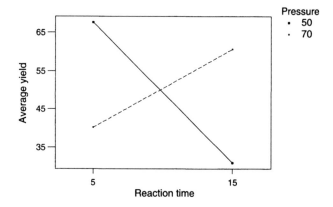

Figure 6.8 Interaction plot – pressure × reaction time.

level (50 psi) and reaction time at high level (15 min). Maximum yield is obtained when the pressure and reaction time are kept at low levels.

6.3.2 Objective 2: To identify the significant main/interaction effects which affect the variability in process yield

In order to identify the significant main/interaction effects which affect process variability, we need to construct a coded design matrix with ln(SD) as the response of interest. Table 6.7 illustrates the design matrix with variability as the response. Due to zero degrees of freedom for the error term, we need to rely on a procedure called 'pooling' of insignificant effects. Pooling is a process of obtaining a more accurate estimate of error variance. Taguchi advocates pooling effects until the degrees of freedom for the error term is approximately equal to half the total degrees of freedom for the experiment.

For the present example, the author has pooled interactions TR, TP and TPR so that three degrees of freedom have been created for the error term. A Pareto plot of the effects is shown in Figure 6.9. The figure shows that none of the main effects have any impact on variability. Interaction between pressure (P)

Table 6.7 Design matrix with variability as response of interest

Run	T	P	R	SD	ln(SD)
1	−1	−1	−1	2.214	0.795
2	1	−1	−1	1.090	0.086
3	−1	1	−1	4.632	1.533
4	1	1	−1	2.940	1.078
5	−1	−1	1	4.707	1.549
6	1	−1	1	4.032	1.394
7	−1	1	1	2.746	1.010
8	1	1	1	2.237	0.805

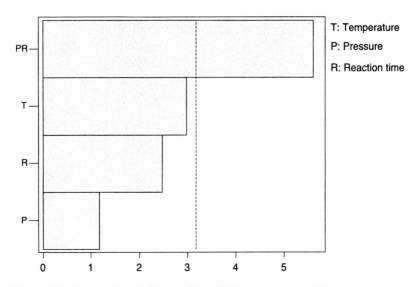

Figure 6.9 Pareto plot of effects with ln(SD) as response of interest.

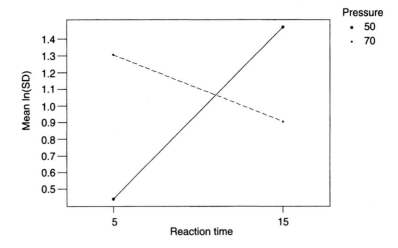

Figure 6.10 Interaction plot of ln(SD) – pressure × reaction time.

and reaction time (R) seems to have some impact on variability (Figure 6.10). It can be seen that variability is minimum when pressure is kept at low level and reaction time at low level.

6.3.3 Objective 3: What is the optimal process condition?

In order to determine the optimal condition of the process, it is important that we need to analyse both response mean and variability. The best settings for maximizing the process yield is:

Temperature (T) – High level (120 °C)
Pressure (P) – Low level (50 psi)
Reaction time (R) – Low level (5 min)

Similarly, the best settings for minimizing response variability is:

Temperature (T) – High level (120 °C)
Pressure (P) – Low level (50 psi)
Reaction time (R) – Low level (5 min)
The above settings can be easily obtained by analysing the mean process
yield and mean ln(SD) values at both low and high level settings of T, P and R.

For normality assumption of data, it is best to construct a NPP of residuals
(Figure 6.11). The graph indicates that the data come from a normal population.

6.4 Example of a 2^4 full factorial design

Here, the author will consider an example with four factors. This example
shows the results of an experiment to study the effect of four factors on
a cracking problem. A nickel–titanium alloy is used to make components for
jet turbine aircraft engines. Cracking is a potentially serious problem in the
final part, because it can lead to non-recoverable failure and subsequent
rejection of the part thereby causing waste. The objective of the experiment
was therefore to identify the key factors and their interactions (if existing)
which have effect on cracks. Four factors were considered (pouring temperature (A), titanium content (B), heat treatment method (C) and the amount of
grain refiner used (D). Each factor was studied at 2-levels and a 2^4 full
factorial design was selected. Table 6.8 presents the experimental layout used
for this experiment to minimize cracks. The response of interest to the

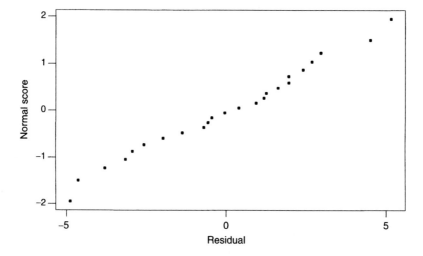

Figure 6.11 Normal probability plot of residuals for the yield experiment.

Table 6.8 Experimental layout with response values

Run	A	B	C	D	Crack length	
1	−1	−1	−1	−1	7.037	6.376
2	1	−1	−1	−1	14.707	15.219
3	−1	1	−1	−1	11.635	12.089
4	1	1	−1	−1	17.273	17.815
5	−1	−1	1	−1	10.403	10.151
6	1	−1	1	−1	4.368	4.098
7	−1	1	1	−1	9.360	9.253
8	1	1	1	−1	13.440	12.923
9	−1	−1	−1	1	8.561	8.951
10	1	−1	−1	1	16.867	17.052
11	−1	1	−1	1	13.876	13.658
12	1	1	−1	1	19.824	19.639
13	−1	−1	1	1	11.846	12.337
14	1	−1	1	1	6.125	5.904
15	−1	1	1	1	11.190	10.935
16	1	1	1	1	15.653	15.053

experimenter was the length of crack (in mm $\times 10^{-2}$). Each trial condition was replicated twice to estimate error variance.

The following are the objectives of the experiment:

1. Which of the main/interaction effects affect mean crack length?
2. Which main effects or interactions might influence variability in crack length?
3. What is the optimal process condition to minimize mean crack length?

6.4.1 Objective 1: Which of the main/interaction effects affect mean crack length?

In order to identify the key main and interaction effects which affect crack length, a Pareto plot of effects (Figure 6.12) was constructed. The Pareto plot clearly indicates that all the main effects (A, B, C and D) and two 2-factor interactions (AB and AC) are statistically significant at 5 per cent significance level. In order to understand the nature of interactions among the factors, the author would suggest the readers to refer to Figure 6.13.

Figure 6.13 Interactions graph for the experiment Figure 6.13 indicates that there is a strong interaction between A and B; and A and C (due to non-parallel lines). We don't generally study three-factor (or three-way) interactions as they are not important in real life settings.

6.4.2 Objective 2: Which of the main/interaction effects affect variability in crack length?

For many industrial experiments, it is important to understand what factors affect mean response and what affect response variability. For optimization

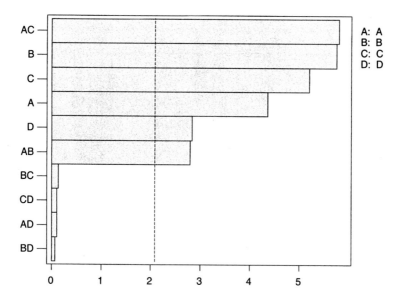

Figure 6.12 Pareto plot of effects for the above example.

problems, we need to minimise response variability around the target performance. This is one of the fundamental objectives of robust design methodology.

In order to analyse what factors affect variability in crack length, we need
to construct a design matrix with ln(SD) as the response. Table 6.9 presents
the design matrix with ln(SD) as the response of interest.

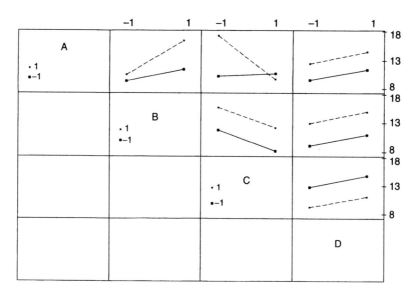

Figure 6.13 Interactions graph for the experiment.

Table 6.9 Experimental layout with response values

Run	A	B	C	D	SD	ln(SD)
1	−1	−1	−1	−1	0.467	−0.761
2	1	−1	−1	−1	0.362	−1.016
3	−1	1	−1	−1	0.321	−1.136
4	1	1	−1	−1	0.383	−0.960
5	−1	−1	1	−1	0.178	−1.726
6	1	−1	1	−1	0.191	−1.655
7	−1	1	1	−1	0.076	−2.577
8	1	1	1	−1	0.366	−1.005
9	−1	−1	−1	1	0.276	−1.287
10	1	−1	−1	1	0.131	−2.033
11	−1	1	−1	1	0.154	−1.871
12	1	1	−1	1	0.131	−2.033
13	−1	−1	1	1	0.347	−1.058
14	1	−1	1	1	0.156	−1.858
15	−1	1	1	1	0.180	−1.715
16	1	1	1	1	0.424	−0.858

In order to identify the factors/interactions which affect variability in crack length, a Pareto plot of effects was constructed (Figure 6.14). The Pareto plot has shown that none of the main effects have significant effect on variability in crack length. Two interactions (AB and CD) are believed to have significant impact on the variability. Figure 6.15 illustrates the interaction plot between factors A and B. It is quite clear from the graph that there exists a strong interaction between the factors A and B. The variability in crack length is minimum when A is kept at low level and B at high level. Similarly, C at low

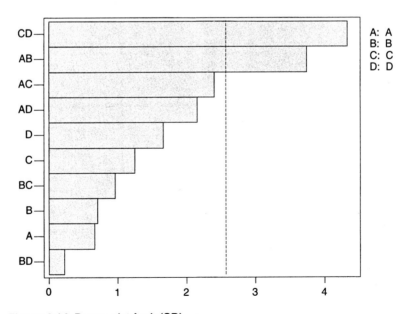

Figure 6.14 Pareto plot for ln(SD).

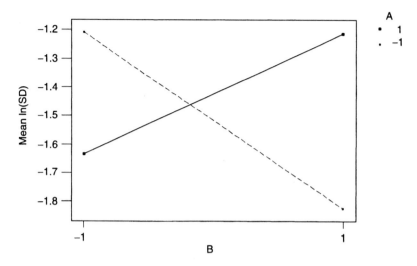

Figure 6.15 Interaction between A and B (response: ln(SD)).

level and D at high level yields minimum variability in crack length. However, it is interesting to observe that factor D is less sensitive to variability when C is kept at high level.

6.4.3 Objective 3: What is the optimal process condition to minimize mean crack length?

In this section, the author will demonstrate how to determine the settings of A, B, C and D to minimize mean crack length. As interactions AB and AC have a significant impact on mean crack length, we need to analyse the mean crack length for all the four combinations between these two factors. Tables 6.10 and 6.11 present the mean crack length at all combinations of factor levels of A and B and A and C respectively.

It is also observed that factor D at low level yields minimum crack length. Therefore the optimal condition of the process to minimize crack length is:

Factor A – Low level (–1)
Factor B – Low level (–1)

Table 6.10 Mean crack length for all combinations of A and B

A	B	Mean crack length
–1	–1	9.458
1	–1	10.542
–1	1	11.5
1	1	16.453

Table 6.11 Mean crack length for all combinations of A and C

A	C	Mean crack length
−1	−1	10.273
1	−1	17.300
−1	1	10.684
1	1	9.696

Factor C – High level (1)
Factor D – Low level (−1)

The NPP of residuals (Figure 6.16) shows that the data come from a normal population.

6.5 Summary

A full factorial experiment assists experimenters to study all possible combinations of the levels of the factors or process parameters in the experiment. By performing a full factorial experiment, one may be able to study the joint effects of two factors (or interactions) on a response by simultaneously changing the levels of factors. This chapter illustrates the use of full factorial designs in industrial experiments and how to analyse and interpret the results of experiments using simple but powerful graphical tools generated by Minitab software system. One of the major limitations of full factorial designs is that the size of the experiment is a function of the number of factors considered and to be studied for the experiment. The rule of thumb therefore is to use a

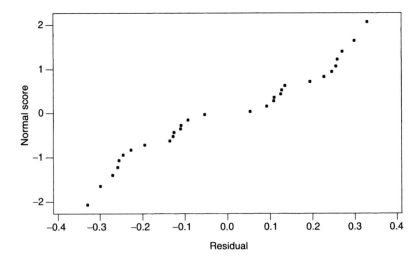

Figure 6.16 Normal probability plot of residuals for the above data.

full factorial design when the number of factors or process parameters is less than or equal to four. When the number of factors is more than four, one may look into fractional factorial designs, which is the focus of next chapter.

Exercises

1. An engineer is interested in the effects of cutting speed (CS), tool geometry (TG), and cutting angle (CA) on the life (in hours) of a machine tool. A 2^3 full factorial design was chosen and the results are shown below. Each trial condition was replicated twice.

Run	CS	TG	CA	Life	
1	−1	−1	−1	22	31
2	1	−1	−1	32	43
3	−1	1	−1	35	34
4	1	1	−1	55	47
5	−1	−1	1	44	45
6	1	−1	1	40	37
7	−1	1	1	60	50
8	1	1	1	39	41

(a) Which effects appear to have significant effect on the tool life?
(b) What is the optimal condition if the objective of the experiment is to maximize tool life?
(c) How do you validate the assumption of normality?

2. In a certain casting process for manufacturing jet engine turbine blades, the objective of the experiment is to determine the most significant main and interaction effects that affect the part shrinkage. Three factors (mould temperature (A), metal temperature (B) and pour speed (C) were studied at two-levels using a 2^3 full factorial experiment. The following table presents the results of the experiment. Each trial condition was replicated three times to obtain sufficient degrees of freedom for the error term.

Run	C	B	A	Shrinkage values (%)		
1	−1	−1	−1	2.22	2.11	2.14
2	1	−1	−1	1.42	1.54	1.05
3	−1	1	−1	2.25	2.31	2.21
4	1	1	−1	1.00	1.38	1.19
5	−1	−1	1	1.73	1.86	1.79
6	1	−1	1	2.71	2.45	2.46
7	−1	1	1	1.84	1.76	1.70
8	1	1	1	2.27	2.69	2.71

(a) Which effects appear to have significant effect on percentage shrinkage?
(b) Which effects appear to have significant effect on variability in shrinkage?
3. A 2^3 full factorial experiment was conducted to study the influence of temperature (A), pressure (B) and cycle time (C) on the occurrence of splay in an injection moulding process. For each of the 8 unique trials, 50 parts were made and the response of interest to the experimenter was the number of incidences of the occurrence of splay on the surface of the part across all 50 parts. The following table shows the experimental layout with the data.

Run	A	B	C	Response
1	−1	−1	−1	12
2	1	−1	−1	15
3	−1	1	−1	24
4	1	1	−1	17
5	−1	−1	1	24
6	1	−1	1	16
7	−1	1	1	24
8	1	1	1	28

(a) Compute all the main and interaction effects.
(b) Construct a Pareto plot of the effect estimates. Which of the effects appear to be statistically significant?

References

Box, G.E.P., Hunter, W.G. and Hunter, J.S. (1978). *Statistics for Experimenters*. NY, John Wiley.

Dean, A. and Voss, D.T. (1999). *Design and Analysis of Experiments*. USA, Springer Verlag.

Logothetis, N. (1992). *Managing for Total Quality*. UK, Prentice-Hall publishers.

Mark J. Kiemele, Stephen R. Schmidt and Ronald J. Berdine (1997). *Basic Statistics: Tools for Continuous Improvement* (4th Edition). Colorado Springs, CO, USA, Air Academy Associates.

Montgomery, D.C. (2001). *Design and Analysis of Experiments* (5th Edition). USA, John Wiley and Sons.

Oehlert, G.W. (2000). *A First Course in Design and Analysis of Experiments*. USA, WH Freeman & Co.

Taguchi, G. (1987). *System of Experimental Design*. NY, Kraus International Publication, UNIPUB.

7

Fractional factorial designs

7.1 Introduction

Very often experimenters do not have adequate time, resources and budget to carry out full factorial experiments. If the experimenters can reasonably assume that certain higher-order interactions (third-order and higher) are not important, then information on the main effects and two-order interactions can be obtained by running only a fraction of the full factorial experiment. A type of orthogonal array design which allows experimenters to study main effects and desired interaction effects in a minimum number of trials or experimental runs is called a fractional factorial design. These fractional factorial designs are the most widely and commonly used types of design in industry. These designs are generally represented in the form $2^{(k-p)}$, where k is the number of factors and $1/2^p$ represents the fraction of the full factorial 2^k. For example, $2^{(5-2)}$ is a 1/4th fraction of a 2^5 full factorial experiment. This means that one may be able to study 5 factors at 2-levels in just 8 experimental trials instead of 32 trials.

7.2 Construction of half-fractional factorial designs

The construction of half-fractions of a full factorial experiment is simple and straightforward. Consider a simple experiment with 3 factors. Table 7.1 shows the design matrix with all the main and interaction effects assigned to various columns of the matrix. Based on our assumption about three-factor (or third-order) and higher-order interactions being negligible, one could use the ABC interaction column in Table 7.1 to generate settings for the fourth factor D. In other words, we would be able to study 4 factors using 8 runs by deliberately aliasing factor D with ABC interaction. This is referred to as a $2^{(4-1)}$ factorial design (Table 7.2).

In the above table, $D = ABC$ implies that main effect D is confounded (or aliased) with a third-order interaction ABC. However a third-order interaction is of no interest to experimenters. The design generator of this design is given by $D = ABC$. We refer to design generator as a word. The defining relation of this design is given by: $D \times D = D^2 = ABCD = I$, where 'I' is the identity

Table 7.1 Design matrix of an 8-run experiment with 3 factors

Run	A	B	AB	C	AC	BC	ABC
1	−1	−1	1	−1	1	1	−1
2	1	−1	−1	−1	−1	1	1
3	−1	1	−1	−1	1	−1	1
4	1	1	1	−1	−1	−1	−1
5	−1	−1	1	1	−1	−1	1
6	1	−1	−1	1	1	−1	−1
7	−1	1	−1	1	−1	1	−1
8	1	1	1	1	1	1	1

Table 7.2 Design matrix of an $2^{(4-1)}$ factorial design

Run	A	B	AB	C	AC	BC	D = ABC
1	−1	−1	1	−1	1	1	−1
2	1	−1	−1	−1	−1	1	1
3	−1	1	−1	−1	1	−1	1
4	1	1	1	−1	−1	−1	−1
5	−1	−1	1	1	−1	−1	1
6	1	−1	−1	1	1	−1	−1
7	−1	1	−1	1	−1	1	−1
8	1	1	1	1	1	1	1

element. Once we know the defining relation of a design, we can then generate the alias structure for that particular design.

In the above experiment, I = ABCD (defining relation). In order to determine the alias of A, we multiply both sides of the defining relation by 'A'. This yields:

$$A \times I = A = A \times ABCD = A^2BCD = BCD, \qquad \text{as} \qquad A^2 = 1.$$

We can now generate aliases of B and C as follows:

$$B \times I = B = ACD$$
$$C \times I = C = ABD$$

Because we are generally interested in two-factor interactions, we can also generate aliases for all two factor interactions as follows:

$$I \times AB = A^2B^2CD = CD$$
$$I \times AC = A^2C^2BD = BD$$
$$I \times BC = B^2C^2AD = AD$$
$$I \times AD = A^2D^2CB = BC$$
$$I \times BD = B^2D^2CA = AC$$
$$I \times CD = C^2D^2AB = AB$$

Similarly, we can generate aliases for three-factor interactions as follows:

$$ABC = A^2B^2C^2D = D$$
$$I \times ABD = A^2B^2D^2C = C$$
$$I \times ACD = A^2C^2D^2B = B$$
$$I \times BCD = B^2C^2D^2A = A$$

Table 7.3 presents the complete aliasing pattern (or confounding pattern) for 4 factors in 8 runs. Minitab software generates the confounding pattern for various types of designs involving up to 15 factors at 2-levels.

For the above design, the resolution is IV (as main effects are confounded with three-factor interactions and two-factor interactions are confounded with other two-factor interactions). In real life situations, certain two-factor interactions may be confounded with other two-factor interactions, and hence we cannot determine which of the two-factor interactions are important to that process. Under such circumstances we may use 'fold-over designs'. Fold-over designs are used to reduce confounding when one or more effects cannot be estimated independently or separately. In other words, the effects are said to be aliased. However fold-over designs are used in resolution III designs to break the links between main effects and two-factor interaction effects. For example, if you fold on one factor, say A, then A and all its two-factor interactions will be free from other main effects and two-factor interactions. If you fold on all factors, then all main effects will be free from each other and from all two-factor interactions.

In a fold-over design, one may perform a second experiment where the factor levels are all the opposite of what they were in the first experiment. That is, interchange the -1s and $+1$s before carrying out the second experiment. However such designs are not recommended when limited time and

Table 7.3 Aliasing pattern for $2^{(4-1)}$ factorial experiment

Effect	Alias
A	BCD
B	ACD
C	ABD
D	ABC
AB	CD
AC	BD
BC	CD
AD	BC
BD	AC
CD	AB
ABC	D
ABD	C
ACD	B
BCD	A

resources are available for industrial designed experiments. Under such circumstances, sound engineering judgements coupled with knowledge in the subject-matter would be of great help to experimenters in separating out the main effects from confounded interaction effects.

7.3 Example of a $2^{(7-4)}$ factorial design

The following section describes an example of a fractional factorial design with resolution III. The example is adapted from Box et al. (1978). This example involves an experiment to study the effect of seven factors at 2-levels using eight trials. The response of interest for the experiment was the time (seconds) taken to climb a hill by a particular person on a bicycle. Table 7.4 illustrates the list of factors and their levels used for the experiment.

Table 7.5 presents the experimental layout with the response values. The runs were performed in random order on eight successive days. This is a $2^{(7-4)}$ factorial design with a design resolution III (i.e. main effects are confounded with two-factor interactions).

Minitab software is used for statistical analysis of data. The first step in the analysis is to identify the most important factors which influence the time to cycle up the hill (seconds). A Pareto plot is constructed to identify the key factors (Figure 7.1). The graph shows that positions of gear (D) and dynamo (B) have a significant effect on the time.

Table 7.4 List of factors and their levels for the experiment

Factors	Labels	Low level	High level
Seat	A	Up	Down
Dynamo	B	Off	On
Handlebars	C	Up	Down
Gear	D	Low	Medium
Raincoat	E	On	Off
Breakfast	F	Yes	No
Tyres	G	Hard	Soft

Table 7.5 Experimental design layout of the experiment

Run	A	B	C	D = AB	E = AC	F = BC	G = ABC	Time to climb hill (sec)
1	−1	−1	−1	1	1	1	−1	69
2	1	−1	−1	−1	−1	1	1	52
3	−1	1	−1	−1	1	−1	1	60
4	1	1	−1	1	−1	−1	−1	83
5	−1	−1	1	1	−1	−1	1	71
6	1	−1	1	−1	1	−1	−1	50
7	−1	1	1	−1	−1	1	−1	59
8	1	1	1	1	1	1	1	88

The design generators of the above design are:

$$D = AB, \quad E = AC, \quad F = BC \quad \text{and} \quad G = ABC$$

Therefore defining relation can be obtained as follows:

$$\begin{aligned} I &= ABD = ACE = BCF = ABCG = BCDE = ACDF = ABEF \\ &= CDG = BEG = AFG = DEF = ADEG = BDFG = ABCDEFG \end{aligned}$$

As we are interested only in main effects and two-factor interactions, the seven main effects and their aliases can be generated in the following manner. As all factors were studied at 2-levels, we estimate only the linear effects of the factors which are confounded with two-factor interactions. For instance, the linear effect of A (l_A) is estimated to be 3.5. However, factor A is confounded with three two-factor interactions such as BD, CE and FG.

$$l_A = 3.5 \rightarrow A + BD + CE + FG$$
$$l_B = 12.0 \rightarrow B + AD + CF + EG$$
$$l_C = 1.0 \rightarrow C + AE + BF + DG$$
$$l_D = 22.5 \rightarrow D + AB + CG + EF$$
$$l_E = 0.50 \rightarrow E + AC + BG + DF$$
$$l_F = 1.0 \rightarrow F + AG + BC + DE$$
$$l_G = 2.5 \rightarrow G + AF + BE + CD$$

As only B and D are two significant effects, we need to analyse them further as D is confounded with B and A, and B is confounded with A and D.

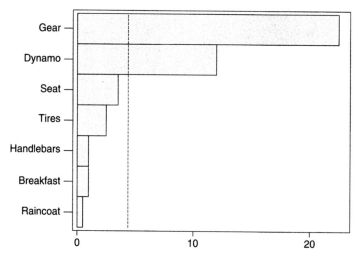

Figure 7.1 Pareto plot of effects for the bicycle data.

Here the largest effect is due to factor D and it is not easy to conclude that the effect of D is large because of just factor D or the confounded two-factor interactions. This problem can be tackled by folding on factor D, by reversing the signs of column containing factor D. This fold-over design is shown in Table 7.6 along with the observed responses. It is quite interesting to observe that both factors B and D are turned out to be significant again (Figure 7.2).

The effects estimated by the second fraction are:

$$l_{A*} = 0.750 \rightarrow A - BD + CE + FG$$
$$l_{B*} = 10.25 \rightarrow B - AD + CF + EG$$
$$l_{C*} = 2.75 \rightarrow C + AE + BF - DG$$
$$l_{D*} = 25.25 \rightarrow D - AB - CG - EF$$
$$l_{E*} = -1.75 \rightarrow E + AC + BG - DF$$
$$l_{F*} = -2.25 \rightarrow F + AG + BC - DE$$
$$l_{G*} = -0.75 \rightarrow G + AF + BE - CD$$

Table 7.6 Fold-over design by folding on just one factor

Run	A	B	C	D = −AB	E = AC	F = BC	G = ABC	Time to climb hill (sec)
1	−1	−1	−1	−1	1	1	−1	47
2	1	−1	−1	1	−1	1	1	74
3	−1	1	−1	1	1	−1	1	84
4	1	1	−1	−1	−1	−1	−1	62
5	−1	−1	1	−1	−1	−1	1	53
6	1	−1	1	1	1	−1	−1	78
7	−1	1	1	1	−1	1	−1	87
8	1	1	1	−1	1	1	1	60

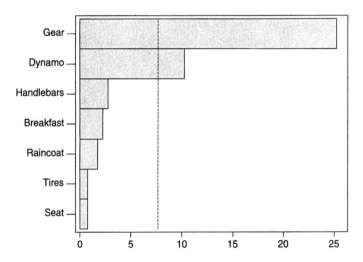

Figure 7.2 Pareto plot of effects for the fold-over design data.

By combining the effect estimates from this second fraction with the effect estimates from the original eight runs, we obtain the following estimates of the effects:

$$l_A + l_{A^*} = 2(A + CE + FG) \quad \text{or} \quad \frac{1}{2}(l_A + l_{A^*}) = A + CE + FG$$

$$\text{i.e. } \frac{1}{2}(3.5 + 0.750) = 2.125 = A + CE + FG$$

Similarly,

$$\frac{1}{2}(10.25 + 12.0) = 11.125 = B + CF + EG$$

$$\frac{1}{2}(2.75 + 1.0) = 1.875 = C + AE + BF$$

$$\frac{1}{2}(22.25 + 22.5) = 23.875 = D$$

$$\frac{1}{2}(-1.75 + 0.5) = -0.625 = E + AC + BG$$

$$\frac{1}{2}(-2.25 + 1.0) = -0.625 = F + AG + BC$$

$$\frac{1}{2}(-0.75 + 2.5) = 0.75 = G + AF + BE$$

We may also write,

$$l_A - l_{A^*} = 2 \times BD \quad \text{or} \quad \frac{1}{2}(l_A - l_{A^*}) = BD$$

i.e.

$$\frac{1}{2}(3.5 - 0.750) = BD \quad \text{or} \quad BD = 1.38$$

Similarly,

$$\frac{1}{2}(12.0 - 10.25) = AD \quad \text{or} \quad AD = 0.88$$

$$\frac{1}{2}(1.0 - 2.75) = DG \quad \text{or} \quad DG = -0.88$$

$$\frac{1}{2}(22.5 - 25.25) = AB + CG + EF \qquad \text{or} \qquad AB + CG + EF = -1.38$$

$$\frac{1}{2}(0.50 + 1.75) = DF = 1.13$$

$$\frac{1}{2}(1.0 + 2.25) = DE = 1.625$$

$$\frac{1}{2}(2.5 + 0.75) = CD = 1.625$$

It can be concluded from the above results that the large main effect due to 'gear' (factor D) is now estimated free of bias from two-factor interactions. The joint effect of three second order interactions (i.e. $AB + EF + CG$) appears to be small. Moreover, all the two-factor interactions involving the factor D are now free of aliases. Similarly, we can conclude that the effect of 2 two-factor interactions (CF and EG) which are aliased with main effect B is shown to be small. Therefore it is safe to say that it is the effect of B which is important in this experiment and has significant impact on the response (i.e. time to climb up the hill).

7.4 An application of 2-level fractional factorial design

In this section, I will now demonstrate another application of a two-level fractional factorial design in the development of a soybean whipped topping. This example is adapted from Chow et al. (1983) published in the *Journal of Food Science*. Non-dairy whipped topping is a fabricated food product that serves as a substitute for whipped cream dessert topping. It is generally formulated with sodium caseinate, vegetable fat, carbohydrates and emulsifiers. The response of interest for this experiment was percentage overrun (or whipability). Seven process variables (or factors) at 2-levels were studied using eight runs. The idea was to separate out the key process variables from the unimportant ones. The experimental layout with responses is shown in Table 7.7. Each trial condition

Table 7.7 Experimental layout for the soybean whipped topping experiment

Run	A	B	C	D = AB	E = AC	F = BC	G = ABC	Overrun (%)
1	−1	−1	−1	1	1	1	−1	115
2	1	−1	−1	−1	−1	1	1	81
3	−1	1	−1	−1	1	−1	1	110
4	1	1	−1	1	−1	−1	−1	69
5	−1	−1	1	1	−1	−1	1	174
6	1	−1	1	−1	1	−1	−1	99
7	−1	1	1	−1	−1	1	−1	80
8	1	1	1	1	1	1	1	63

was randomized to minimize the effect of any noise (or hidden variables) induced into the experiment.

Figure 7.3 presents the main effects plot for the experiment. Main effects A, B, F and G appear to be important whereas main effects due to C, D and E do not appear to be important to the process. These effects have been pooled to generate adequate degrees of freedom for the error term. Figure 7.4 illustrates the Pareto plot of effects which implies that factors A (Soybean emulsion), B (Vegetable fat) and F (Carbohydrates) are statistically significant and therefore should be studied in detail. The next section will look into the design generators, defining relation and confounding or aliasing pattern for the experiment.

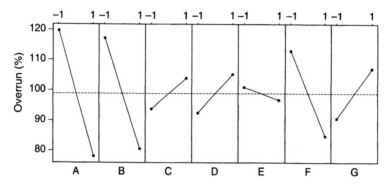

Figure 7.3 Main effects plot for the soybean whipped topping experiment.

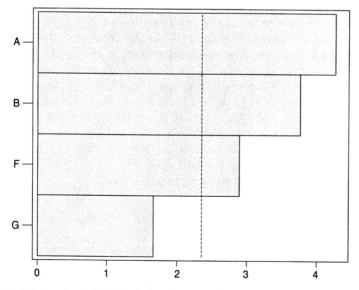

Figure 7.4 Pareto plot of effects for the experiment.

The design generators of the design are:

$$D = AB, \quad E = AC, \quad F = BC \quad and \quad G = ABC$$

The defining relationship for this design is therefore obtained by adding to the generators all of their products taken two, three and four at a time. The complete defining relation is therefore generated as:

$$I = ABD = ACE = BCF = ABCG = BCDE = ACDF = CDG = ABEF$$
$$= AFG = BEG = DEF = CEFG = ADEG = BDFG = ABCDEFG$$

Based on the above defining relations, one can generate the following linear combinations of confounded effects.

$$l_A = -41.75 \rightarrow A + BD + CE + FG$$
$$l_B = -36.75 \rightarrow B + AD + CF + EG$$
$$l_C = 10.25 \rightarrow C + AE + BF + DG$$
$$l_D = 12.75 \rightarrow D + AB + CG + EF$$
$$l_E = -4.25 \rightarrow E + AC + BG + DF$$
$$l_F = -28.25 \rightarrow F + AG + BC + DE$$
$$l_G = 16.25 \rightarrow G + AF + BE + CD$$

From the Pareto plot, we might conclude that the three main effects (A, B and F) are the important variables which affect whipability. But we cannot make any valid conclusions at this point as the main effects due to A, B and F are confounded with a number of two-factor interactions. For example, we cannot conclude that factor A is significant due to its true effect on whipability, rather it is significant due to interactions BD/CE or FG. In order to remove the ambiguity surrounding the results of this experiment, one could perform a fold-over (or mirror image) design. In this case, we have folded on all factors in order to make the main effects free from each other and two-factor interactions. Therefore a second $2^{(7-4)}$ fractional factorial design is performed by switching the signs from -1 to 1 and vice versa for all of the columns in the original experimental layout shown in Table 7.7. The results of the fold-over experiment are shown in Table 7.8.

The design generators of the second fraction are:

$$D = -AB, \quad E = -AC, \quad F = -BC \quad and \quad G = ABC$$

The defining relationship for the folded design is therefore obtained by adding to the generators all of their products taken two, three and four at a time. The complete defining relation for the folded (or mirror image) design is therefore generated as:

Table 7.8 Experimental layout for the soybean whipped topping experiment

Run	A	B	C	D = −AB	E = −AC	F = −BC	G = ABC	Overrun (%)
1	1	1	1	−1	−1	−1	1	84
2	−1	1	1	1	1	−1	−1	69
3	1	−1	1	1	−1	1	−1	56
4	−1	−1	1	−1	1	1	1	161
5	1	1	−1	−1	1	1	−1	56
6	−1	1	−1	1	−1	1	1	40
7	1	−1	−1	1	1	−1	1	92
8	−1	−1	−1	−1	−1	−1	−1	208

$$I = -ABD = -ACE = -BCF = ABCG = BCDE = ACDF = -CDG$$
$$= ABEF = -AFG = -BEF = -DEF = CEFG = ADEF = BDFG$$
$$= -ABCDEFG$$

Based on the above defining relations, one can generate the following linear combinations of confounded effects (assuming that third and higher-order interactions can be neglected).

$$l_{A*} = -47.5 \rightarrow -A - BD - CE - FG$$

$$l_{B*} = -67.00 \rightarrow B - AD - CF - EG$$

$$l_{C*} = -6.50 \rightarrow C - AE - BF - DG$$

$$l_{D*} = 63.00 \rightarrow AB - D + CG + EF$$

$$l_{E*} = 2.50 \rightarrow AC - E + BG + DF$$

$$l_{F*} = 35.00 \rightarrow BC - F + AG + DE$$

$$l_{G*} = -3.00 \rightarrow G - AF - BE - CD$$

By combining the effect estimates from this second fraction with the effect estimates from the original eight runs, we obtain the following estimates of the effects:

$$l_A + l_A = 2A \quad \text{or} \quad \frac{1}{2}(l_A + l_{A*}) = A$$

i.e.

$$\frac{1}{2}(-41.75 + 47.5) = -44.625 = A$$

Similarly,

$$\frac{1}{2}(-67.0 + -36.75) = -51.875 = B$$

$$\frac{1}{2}(10.25 + -6.50) = 1.875 = C$$

$$\frac{1}{2}(12.75 + 63) = 37.875 = (AB + CG + EF)$$

$$\frac{1}{2}(2.50 - 4.25) = -0.875 = (AC + BG + DF)$$

$$\frac{1}{2}(35.00 - 28.25) = 3.375 = (BC + AG + DE)$$

$$\frac{1}{2}(16.25 - 3.00) = 6.625 = G$$

Similarly,

$$\frac{1}{2}(-41.750 - (-47.50)) = 2.875 = BD + CE + FG$$

$$\frac{1}{2}(-36.750 - (-67.00)) = 15.125 = AD + CF + EG$$

$$\frac{1}{2}(10.25 - (-60.50)) = 8.375 = AE + BF + DG$$

$$\frac{1}{2}(12.75 - 63.00) = -25.125 = D$$

$$\frac{1}{2}(-4.25 - 2.50) = -3.375 = E$$

$$\frac{1}{2}(-28.25 - 35.00) = -31.625 = F$$

$$\frac{1}{2}(16.25 - (-3.00)) = 9.625 = AF + BE + CD$$

The estimates of the main effects and sets of three two-factor interactions are summarized in Table 7.9.

An examination of Table 7.9 shows that main effects A, B, D, F and the linear combination of three two-factor interactions (AB, CG and EF) appear to be important. However we cannot tell which of the above three-factor interactions is responsible. It is clear from Table 7.9 that factors C, E and G have no impact on the percentage overrun. Hence it can be concluded that it is

Table 7.9 Estimates of effects from combined designs

Estimate of effect A = −44.625
Estimate of effect B = −51.875
Estimate of effect C = 1.875
Estimate of effect D = −25.125
Estimate of effect E = −3.375
Estimate of effect F = −31.625
Estimate of effect G = 6.625

Estimate of AB + CG + EF = 37.875
Estimate of AC + BG + DF = −0.875
Estimate of BC + AG + DE = 3.375
Estimate of BD + CE + FG = 2.875
Estimate of AD + CF + EG = 15.125
Estimate of AE + BF + DG = 8.375
Estimate of AF + BE + CD = 9.625

AB interaction which is important with respect to the overrun, as both factors A and B have a significant influence on the overrun. Figure 7.5 illustrates the interaction graph between A and B. The graph shows that there exists a strong interaction between A and B.

7.5 Example of a $2^{(5-1)}$ factorial design

The next example is about the investigation of the effect of five factors on the free height of leaf springs used in an automotive application (for more information on the case study, the readers may refer to the *Journal of Quality Technology*, Vol. 17, pp. 198–206, 1985). Table 7.10 presents the

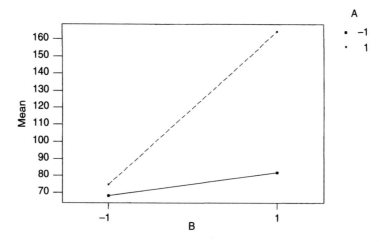

Figure 7.5 Interaction between A and B.

Table 7.10 Experimental layout with response values

Run	A	B	C	D	E	Free height values		
1	−1	−1	−1	−1	−1	7.78	7.81	7.78
2	1	−1	−1	1	−1	8.15	7.88	8.18
3	−1	1	−1	1	−1	7.50	7.56	7.50
4	1	1	−1	−1	−1	7.59	7.75	7.56
5	−1	−1	1	1	−1	7.54	8.00	7.88
6	1	−1	1	−1	−1	7.69	8.06	8.09
7	−1	1	1	−1	−1	7.44	7.52	7.56
8	1	1	1	1	−1	7.56	7.69	7.81
9	−1	−1	−1	−1	1	7.50	7.25	7.12
10	1	−1	−1	1	1	7.44	7.88	7.88
11	−1	1	−1	1	1	7.50	7.56	7.50
12	1	1	−1	−1	1	7.56	7.63	7.75
13	−1	−1	1	1	1	7.32	7.44	7.44
14	1	−1	1	−1	1	7.69	7.56	7.62
15	−1	1	1	−1	1	7.18	7.25	7.18
16	1	1	1	1	1	7.50	7.81	7.59

experimental layout and the recorded values of free height. Each trial condition was replicated three times to determine the variability within the trial conditions. The five factors used for the experiment are: A = furnace temperature, B = heating time, C = transfer time, D = hold down time and E = quench oil temperature. This is a $2^{(5−1)}$ fractional factorial design with design generator D = ABC. In other words, the design resolution of the experiment is IV. This implies that main effects are confounded with 3-factor interactions or 2-factor interactions are confounded with other 2-factor interactions.

The defining relation is given by I = ABCD. The aliasing or confounding structure is shown below.

$$A = BCD, \qquad B = ACD, \qquad C = ABD, \qquad D = ABC$$
$$AB = CD, \qquad AC = BD, \qquad AD = BC$$
$$ABC = D, \qquad ABC = C, \qquad ACD = B, \qquad BCD = A$$

The following are the objectives of this experiment.

1. What factors influence the mean free height?
2. What factors affect variability in the free height of springs?

7.5.1 Objective 1: To identify the factors which influence the mean free height

Minitab software is used to identify the factors which influence the mean free height of leaf springs. Figure 7.6 illustrates a Pareto plot of effects which indicates that main effects A, B, D, E and a 2-factor interaction BE are considered to have significant impact on mean height at 5 per cent significance level. In order to validate the assumption of normality, the author has constructed a normal probability of residuals (Figure 7.7). The normal plot

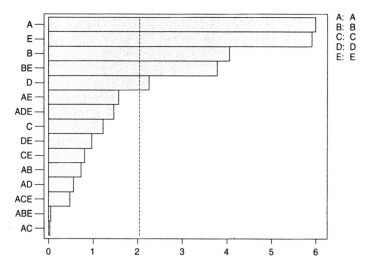

Figure 7.6 Pareto plot of effects for the leaf spring experiment.

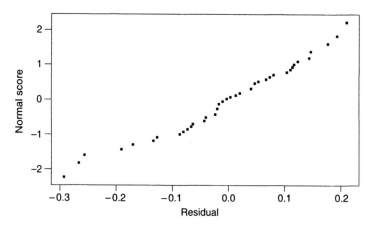

Figure 7.7 Normal probability plot of residuals for the leaf spring example.

has shown that the residuals fall approximately along a straight line and hence we may conclude that the data come from a normal population.

7.5.2 Objective 2: To identify the factors which affect variability in the free height of leaf springs

In order to determine which of the factors or interaction effects have a significant influence on the variability, it was decided to construct a Pareto plot of effects (Figure 7.8). Due to insufficient degrees of freedom for the error term, it was decided to pool the effects with low magnitude.

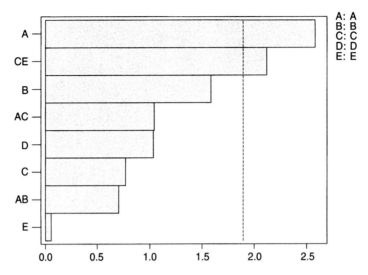

Figure 7.8 Pareto plot of effects which influence variability.

The Pareto plot has indicated that main effect A and interaction effect CE appear to have a significant impact on variability at 10 per cent significance level. The interaction plot (Figure 7.9) implies that there is a strong interaction between the factors C (transfer time) and E (quench oil temperature). It can be observed from the plot that variability in the free height of leaf springs is minimum when both C and E are kept at low levels. Moreover, it can be seen that variability is high when E is kept at low level and C at high level. As main effect C is confounded with a third-order interaction, it is fair

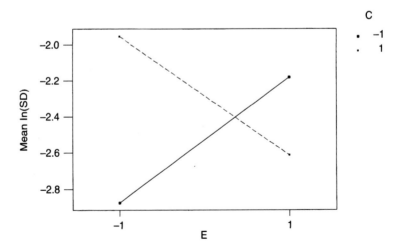

Figure 7.9 Interaction plot between quench oil temperature and transfer time.

to conclude that it is the interaction CE which causes variability in the free height of leaf springs.

7.5.3 How do we select the optimal factor settings to minimize variability in free height?

For any process optimisation problems, it is important to determine the optimal factor settings which meets the experimental objectives. Here we need to determine the best factor settings which yields minimum variability in the free height of leaf springs. A cube plot was constructed with factors A, C and E (Figure 7.10). The cube plot clearly shows that minimum variability is obtained when all the factors are kept at low levels. It can be concluded that the optimal settings for minimizing variability is (Figure 7.11):

Factor A – Low level (-1),   Factor B – High level (1),   Factor C – Low level (-1),   Factor D – Low level (-1),   Factor E – Low level (-1)

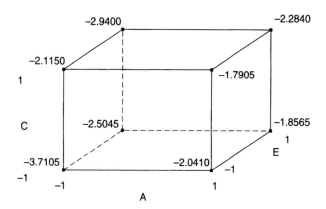

Figure 7.10 Cube plot of effects.

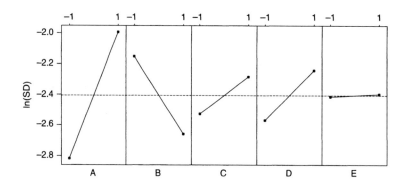

Figure 7.11 Main effects plot for ln(SD).

7.6 Summary

Experimenters utilize fractional factorial designs to study the most important factors or process/design parameters which influence critical quality characteristics. Fractional factorial experiments are used for pilot studies, screening experiments, etc. where the goal is to gain maximum information about a process in a limited number of experimental trials. This chapter provides details for constructing fractional factorial experiments and highlighting the problems associated with highly fractionated factorial experiments wherein main effects are confounded or aliased with two-order interactions. Extensive graphical tools have been used through the use of real-world examples in manufacturing industry. All the graphs were generated using Minitab software system. More real industrial case studies involving fractional factorial experiments are illustrated in the next chapter.

Exercises

1. A $2^{(7-4)}$ fractional factorial design was conducted on a chemical process to evaluate the effect of seven process variables which might influence the yield (%) of the process. The list of variables and their levels used for the experiment are shown below:

Variable	Low level	High level
Temperature (A)	150	200
Pressure (B)	Low	High
Concentration of chemical A (C)	3%	5%
Concentration of chemical B (D)	2%	8%
Type of catalyst (E)	A	B
Reaction time (F)	Low	High
Flow rate (G)	Low	High

Source: DeVor, R.E., Chang, T.-H. and Sutherland, J.W. (1992). *Statistical Quality Design and Control*, NY, Macmillan Publishing Company.

The results of the experiment are shown below. The response for the experiment is per cent yield. Note that the tests are displayed in the order in which they were carried out.

Run	A	B	C	D	E	F	G	Yield (%)
1	−1	1	1	−1	−1	1	−1	66.1
2	−1	1	−1	1	−1	−1	1	59.6
3	1	−1	1	1	−1	−1	−1	62.3
4	1	1	−1	−1	1	−1	−1	67.1
5	−1	−1	1	−1	1	−1	1	21.1
6	1	1	1	1	1	1	1	57.8
7	−1	−1	−1	1	1	1	−1	59.7
8	1	−1	−1	−1	−1	1	1	22.5

(a) What are the generators and defining relation for this experiment?
(b) Illustrate the complete confounding structure for the design, assuming third-order and higher-order interactions are negligible.
(c) Which factor or interaction effects appear to have significant impact on percentage yield?
(d) Construct a Pareto plot of effects and determine the optimal settings of the variables which gives maximum yield.
(e) How do you validate the assumption of normality?

2. An experimenter decided to study the effect of four process parameters for an injection moulding process. The experimenter was interested in both main and two-factor interactions. The response of interest was the width of the injected part (accuracy is up to four decimal places), which is critical to customers. The results of the experiment are shown in the following table. The experiment was repeated twice to create sufficient degrees of freedom for the error term. The four process variables are: (1) D: Mould temperature, (2) A: Injection speed, (3) E: Hold pressure and (4) B: Cooling time.

Trial no.	D	A	E	B = DAE	Width	
1	−1	−1	−1	−1	9.3415	9.3416
2	−1	−1	1	1	9.3691	9.3692
3	−1	1	−1	−1	9.3467	9.3466
4	−1	1	1	−1	9.3680	9.3681
5	1	−1	−1	1	9.3679	9.3680
6	1	−1	1	−1	9.3493	9.3494
7	1	1	−1	1	9.3668	9.3669
8	1	1	1	1	9.3544	9.3545

Source: Schmidt, S.R. and Launsby, R.G. (1992). *Understanding Industrial Designed Experiments*, Colorado Springs, Colorado, Air Academy Press.

(a) What is the resolution of this design?
(b) Display the complete confounding structure.
(c) Which effects appear to have significant effect on the width?
(d) What are the best settings of the parameters to achieve a target width of 9.380?

3. An experimenter is interested to study the effect of five welding process parameters. The results of the experiment are illustrated below. The response of interest to the experimenter is heat input (measured in watts) for welding. The welding parameters considered for the experiment are: A: open-circuit voltage, B: slope, C: electrode melt-off rate, D: electrode diameter and E: electrode extension.

Source: Stengner, D.A.J. et al. (1967) Prediction of heat input for welding, Welding, J. res. Supplement, 1, March.

The design matrix of the experiment with response is shown in the following table.

Trial no.	A	B	C	D	E	Heat input (W)
1 (12)	−1	−1	−1	−1	1	3318
2 (1)	1	−1	−1	−1	−1	4141
3 (2)	−1	1	−1	−1	−1	3790
4 (6)	1	1	−1	−1	1	4061
5 (15)	−1	−1	1	−1	−1	3431
6 (8)	1	−1	1	−1	1	3425
7 (7)	−1	1	1	−1	1	3507
8 (4)	1	1	1	−1	−1	3765
9 (11)	−1	−1	−1	1	−1	2580
10 (14)	1	−1	−1	1	1	2450
11 (3)	−1	1	−1	1	1	2319
12 (16)	1	1	−1	1	−1	3067
13 (13)	−1	−1	1	1	1	1925
14 (10)	1	−1	1	1	−1	2466
15 (5)	−1	1	1	1	−1	2485
16 (9)	1	1	1	1	1	2450

Note: ( ) implies the order in which the experimental trials were carried out.

(a) What is the defining relation of this design?
(b) Display the complete confounding structure and determine the design resolution.
(c) Which effects appear to have significant effect on heat input?
(d) Construct a normal probability plot of residuals for validating normality assumptions.

References

Bisgaard, S. (1988). A *Practical Aid for Experimenters*. Madison, Wisconsin, Starlight Press.

Box, G.E.P., Hunter, W.G. and Hunter, J.S. (1978). *Statistics for Experimenters*. NY, John Wiley.

Box, G.E.P. (1992). What can you find out from eight experimental runs? *Quality Engineering*, 4(4), 619–627.

Chow, E.T.S., Wei, L.S., DeVor, R.E. and Steinberg, M.P. (1983). Application of a two-level fractional factorial design in the Development of a Soybean Whipped Topping, *Journal of Food Science*, pp. 230.

Daniel, C. (1976). *Applications of Statistics to Industrial Experimentation*. NY, John Wiley.

Drain, D. (1997). *Handbook of Experimental Methods for Process Improvement*. London, UK, Chapman and Hall.

Taguchi, G. (1986). *Introduction to Quality Engineering, Quality Resources*. NY, White Plains.

8

Some useful and practical tips for making your industrial experiments successful

8.1 Introduction

Experimental Design (ED) is a powerful approach to achieve increased understanding of your process, leading to significant improvements in product quality, decreased manufacturing costs and potentially thousands of dollars of savings for organizations. So why don't more manufacturers use ED? Why do some manufacturing companies try ED, and then abandon it saying 'it won't work for us'? Inadequate training, demanding production schedules or time pressures, cost and resources required for the execution of an experiment or a series of experiments often cited as principal reasons. Moreover, fear of statistics is widespread, even among many educated scientists and managers in organizations. This chapter provides some useful and practical tips to industrial engineers and managers with limited knowledge in ED for making industrial experiments successful in their own organizations. The purpose of this chapter is to stimulate the engineering community to start applying ED for tackling quality control problems in key processes they deal with everyday.

Industrial experiments are fundamental and crucial to increase the understanding of a process and product behaviour. The success of any industrial experiment depends on a number of key factors such as statistical skills, engineering skills, planning skills, communication skills, teamwork skills and so on. Many scientists and engineers perform industrial experiments based on full and fractional factorial designs (Montgomery, 1991) or Orthogonal Array (OA) designs (Taguchi, 1986) for improving the product quality and process efficiency. In other words, engineers and managers of today's modern industrial world have placed an increased emphasis on achieving breakthrough improvements in product and process quality using ED.

Experimental Design is essentially a strategy of industrial experimentation whereby one may vary a number of factors in a process/system simultaneously to study their effect on the process/system output. It is a direct replacement of traditional One-Factor-At-A-Time (OFAT) or 'Hit or Miss' approach to experimentation (Antony, 1998). It is important to note that these tips were developed strictly on the basis of author's experience and expertise in the field of study and also by reviewing many industrial case studies and literature in the subject matter.

8.1.1 Get a clear understanding of the problem

One of the key reasons for an industrial experiment to be unsuccessful is lack of understanding of the problem itself. The nature of the experiment to be conducted is heavily dependent on the nature of the problem and objective of the experiment. Therefore, it is absolutely essential to have a clear definition of the problem and the objective of the experiment before one embarks on to any kind of experimentation. A well-defined objective leads the experimenter to the correct choice of ED. If you incorrectly state the objective(s) of an experiment, you may have to face the consequences – trying to study too many or too few factors for the experiment, not measuring the right quality characteristics or responses, or arriving at conclusions which are already known to the team conducting the experiment. In other words, unclear objectives would lead to lost time and money, as well as lack of appreciation and feelings of frustration for all involved in the study. Industrial experiments are generally a team effort; a typical team includes people from design, quality and production department and an operator. It is quite important that everyone in the team should have a clear understanding of the objective of the experiment and also their role in experimentation. If there is more than one objective, it is then important to assign a relative weight to the objectives and establish ways that each will be evaluated.

8.1.2 Project selection

Selection of the right project may assure you a success or guarantee an opportunity to try it again the second time. Many companies are continuously engaged in a number of ED projects and it is important to identify the projects that can return the most savings. In situations where you have a number of experiments to be performed for a variety of problems, it is worthwhile considering the following factors in mind.

(a) *Management involvement and commitment*. Management must be involved in the project right from the beginning. You need their support and commitment, when you need to take necessary improvement actions on the process or system.

(b) *Return on investment*. When you have a number of experiments to be carried out, consider the return on investment. Savings from reduced warranty costs, reduced customer complaints and increased customer satisfaction may produce a higher return in the long term.

(c) *Project scope.* If the system or process you deal with for experimentation purposes is too intricate in nature, it is best to break it down into sub-systems or sub-processes. For example, rather than optimizing the entire vehicle, it is better to start optimizing the braking or suspension system. If it is feasible and practical, you may break the braking system into many subsystems and seek to optimize the surface finish of the rotor disk.

(d) *Time required to complete the project.* An unfinished experiment is a waste of time and resources and it can be quite detrimental to all future initiatives. So it is important to start off with projects which bring quick wins to the organization in a short time. This would help to boost the morale of the team and assist them to become more confident to undertake more and more projects across the organization.

(e) *Value to your organization.* You should select a project that adds long-term value to the future of your organization. Carry out projects (in the form of experiments) to achieve greater product performance that your customers are not asking for now but may ask for soon.

8.1.3 Conduct exhaustive and detailed brainstorming session

Many DOE training courses and text books might spend as much as 70–80 per cent of their time in the analysis of experiment (i.e. statistical skills). The successful application of ED in today's industrial environment requires a mixture of statistical, planning, engineering, communication and teamwork skills. Brainstorming must be treated as an integral part in the design of effective experiments. There is no standard procedure on how to perform a typical brainstorming as applicable to all industrial situations. The nature and content of each brainstorming session will rely heavily on the nature of the problem under investigation. In the context of ED, brainstorming is performed with the following purposes in mind:

- Identification of the factors, the number of levels and other relevant information about the experiment.
- Development of team spirit and positive attitude in order to assure greater participation of the team members.
- How well does the experiment simulate the customers or users conditions?
- Who will do what and how? For example, who will be responsible for data analysis?
- How quickly does the experimenter need to provide the results to the management?
- Is experimentation the only way to tackle the problem at hand?

8.1.4 Teamwork and selection of a team for experimentation

For ED projects, it is good practice to have a project owner who is responsible for team formation. In selecting team members, the following criteria may be considered:

- *Project beneficiaries* – These are people who must accept your recommendation for improvement further to key findings from the experiment. They may not be directly involved in the project, however it is important to bring them in the loop somehow.
- *Parts/material supplier* – If the supplier of the material is a factor for the experiment, it is best to consult with the suppliers and it will be good to include them in the experimentation team.
- *Direct involvement* – When planning and conducting an experiment, it is important to include people who can provide input in the identification of factors for the experiment. For a typical industrial designed experiment, personnel involved in design, validation, quality, production, and operators are likely candidates.

8.1.5 Select the continuous measurable quality characteristics or responses for the experiment

A quality characteristic or response is the performance characteristic of a product which is most critical to customers and often reflects the product quality. Selecting the right quality characteristic (or response) is critical to the success of any industrial designed experiment. Many DOE programs fail because their responses cannot be measured quantitatively. A classic example can be found with traditional approach to evaluate quality, where an inspector uses a subjective judgement based on his experience to determine whether a product or unit passes or fails the test. Pass/fail data can be used in DOE, but it is very crude and inefficient. For example, if your process typically produces a 0.5 per cent defect rate, you would expect to find 5 out of 1000 parts defective. If you perform a 16 trial experiment, you would then require a minimum of 16,000 parts (16×1000). This poses the question 'Can we afford the cost associated with the parts?'

The following guidelines may be useful to engineers in selecting the quality characteristics or responses for industrial experiments.

- Use quality characteristics (or responses) that can be measured accurately and with stability.
- Use quality characteristics that can be measured quantitatively.
- Use quality characteristics which are directly related to the energy transfer associated with the fundamental mechanism of the product or the process.
- Use quality characteristics which are complete, i.e. they should cover the input–output relationship for the product or the process.

- For complex systems or processes, select quality characteristics at the sub-system level and perform experiments at this level before trying to optimize the overall system.

Consider a coating process which results in various problems such as poor appearance, low yield, orange peel and voids. Too often, experimenters measure these characteristics as data and try to optimize the response. This is not sound engineering, because these are the symptoms of poor function. It is not the function of the coating process to produce an orange peel. The problems such as orange peel are due to excessive variability of the coating process caused by noise factors such as variability in viscosity, ambient temperature, etc. We should measure data that relate to the function itself, not the symptom of variability. One fairly good characteristic to measure for the coating process is the coating thickness. The aim of the coating process is to form the coating layer; effects such as orange peel result from excessive variability of coating thickness from its target. A sound engineering approach is to measure the coating thickness and determine the best settings of the coating process that will minimize the coating thickness variability around its target value. Table 8.1 provides a framework covering a variety of manufacturing process problems, and the suitable response of interest to experimenters for each associated process.

In essence, the selection of attribute quality characteristics (e.g.: good/bad, defective/non-defective, etc.) for industrial experiments is not a good practice. This does not mean that experimenters should measure only continuous measurable quality characteristics. The author nevertheless recommends to choose continuous characteristics over attributes. One of the limitations with the attribute characteristic is its poor additivity property. It means that many main effects will be confounded with two-factor interactions or two-factor interactions will be confounded with other two-factor interactions. Attribute characteristics also require a large number of samples and therefore experiments involving such characteristics are costly and time consuming.

Table 8.1 Examples of quality characteristics for various manufacturing processes

Type of process	Objective of the experiment	Appropriate response
Extrusion	To reduce the post extrusion shrinkage of a speedometer cable casing	Shrinkage
Coil Spring manufacturing	To reduce variability in the tension of coil springs	Spring tension
TV picture tube manufacturing	To reduce performance variation of TV electron guns	Cut-off voltage
Surface mounting	To improve field reliability	Shear strength
Gold plating	To reduce variation in gold plating thickness	Plating thickness
Die-casting process	To increase the hardness of a die-case engine component	Rockwell hardness
MIG welding	To reduce the high scrap rate due to poor welded joints	Weld strength
Wire bonding	To reduce the defect rate from broken wires	Wire pull strength

8.1.6 Choice of an appropriate Experimental Design

The choice of ED is very important for the success of any industrial experiment as it depends on a various number of factors which include the nature of the problem at hand, the number of factors to be studied, resources available for the experiment, time needed to complete the experiment and the resolution of the design. We can use either classical experimental design (full and fractional factorial designs) advocated by Sir Ronald Fisher or Orthogonal Array (OA) designs recommended by Dr Taguchi. For classical experimental design, the focus is on the study of product and process behaviour, followed by the development of a mathematical model which explicitly illustrates the relationship between a dependent variable and a set of independent variables. Experiments based on OA designs promoted by Taguchi are focused on product and process robustness. Here robustness refers to reducing the process/product performance to noise sensitivity. Taguchi recommends the use of the Signal-to-Noise Ratio (SNR) to estimate the performance sensitivity of a product to noise. The choice of any of these designs will be dependent upon the following factors:

- degree of optimization required for the chosen quality characteristic
- number of factors and interactions (if any) to be studied
- complexity of using each design
- statistical validity and effectiveness of each design
- degree of product/process functional performance robustness to be attained from the experiment
- ease of understanding and implementation
- nature of the problem (or objective of the experiment)
- cost and time constraints.

8.1.7 Iterative experimentation

Experiments should be conducted in an iterative manner so that information gained from one experiment can be applied to the next. It is best to run a number of smaller and sequential experiments rather than running a large experiment with several factors and using up the majority of resources assigned to the experimentation process. If none of the factors or process variables are significant, the experiment would then be a waste of time and money. The first step in any experimentation process is to 'separate out the vital few from the trivial many'. Screening experiments are generally performed to reduce the number of factors or key process variables to a manageable number in a limited number of experimental trials.

It is advisable not to invest more than 25 per cent of the experimental budget in the first phase of any experimentation such as screening. Having identified the key factors, the interactions among them can be studied using full or fractional factorial experiments. Once you identify the key variables and interactions for a process, you may then want to perform a Response

Surface Methodology (RSM), which allows you to model the process behaviour over its entire operating region. Using RSM, one may be able to develop a second-order mathematical model that depicts the relationship between the key process variables and the process response. This model can then be used to predict the values of the responses at different variable settings.

8.1.8 Randomize the experimental trial order

In the context of ED, randomization is a process of performing experimental trials in a random order in which they are logically listed. This is a very important concept in any ED because an experimenter cannot always be certain that all important factors affecting a response have been included and considered in the experiment. The purpose of randomization is to reduce the systematic bias that is induced into the experiment. The bias may be due to the effect of uncontrolled factors or noise, such as machine ageing, changes in raw material, tool wear, change of relative humidity, power surges, change of ambient temperature and so on. These changes, which often are time-related, can significantly influence the response. For example, assume that an experiment is performed so that all the low levels of factor A are run first, followed by the high levels of factor A. During the course of the experiment, the humidity in the work place changes by 50 per cent, creating a significant effect on the response. The analysis may reveal that factor A is statistically significant. In reality factor A is not significant, it is the change in humidity level that caused the factor effect to be significant. Randomization would have prevented this confusion.

Whilst conducting an experiment, do not underestimate the background noise inherent in the experiment. Characterization of the noise variables allows an engineer the ability to understand their effect and minimize their influence on the process performance. A factor may turn out to be significant due to the influence of the lurking variables (or noise variables) which often are uncontrollable. Randomization will minimize the effect of a factor which has been confounded with the effect of noise. The author therefore recommends the experimenters to randomize (if possible) the trials.

8.1.9 Replicate to dampen the effect of noise or uncontrolled variation

Replication improves the chance of detecting a statistically significant effect (i.e. signal) in the midst of natural process variation. In some processes, the amount of natural process variation is very large. This can mitigate your chance to detect a significant factor or interaction effect. One of the common queries before conducting experiments in organizations is 'how many experimental runs are required to identify significant effect(s), given the current process variation?'. Signal-to-Noise ratios help one determine the minimum experimental runs needed to achieve a given power for your ED. The signal is the change in response that you want to detect. You need to determine the smallest change you want to detect. Once the signal is detected, you may then estimate the noise. Here noise is the random variation that occurs in the response during standard operating conditions. The noise (i.e. measure of variation) can be estimated from either control charts (using the equation: $\sigma = d_2/R$ or the ANOVA table from a designed experiment (refer to the value of Root Mean Square Error: RMSE).

The number of replications is a direct function of the size of the experiment. Table 8.2 provides some guidance to determine how many experimental runs are required to be conducted for the desired detectable signal. If you cannot afford to perform the necessary runs, then you must find some ways to minimize the noise or random variation. The number of runs is given by the following formula:

$$N = \frac{(4r)^2}{(\Delta/\sigma)^2} \qquad (8.1)$$

where N is total number of experiments, r is the number of levels of the factors, Δ is the size of the effect to detect and σ is the noise level. The derivation of the above equation is based on providing approximately a 90 per cent confidence of finding an active effect of size Δ. For example, for an injection moulding process, the management would like to reduce the shrinkage by 0.85 per cent (i.e. $\Delta = 0.85$). The SD of the process is known to be about 0.60 per cent (i.e. $\sigma = 0.60$). Assume that each factor is studied at 2-levels. The total number of experiments in this case can be computed (using Eq. (8.1) as 32. Consider another example where the objective of the experiment is to improve the yield of a chemical process by 1 per cent. The SD of the process is estimated to be 0.5 per cent. The minimum number of experiments to detect an effect of 1 per cent is 16.

Many process engineers engaged in industrial experiments are not sure of the difference between repetition and replication. Replication is a process of running the experimental trials in a random fashion. In contrast, repetition is a process of running the experimental trials under the same set up of machine parameters. In other words, the variation due to machine set-up cannot be captured using repetition. Replication requires resetting of each trial condition

Table 8.2 Number of experiments as a function of signal-to-noise ratio

Signal-to-noise ratio (Δ/σ)	Minimum number of experiments
1.0	64
1.4	32
2.0	16
2.8	8

and therefore the cost of the experiment and also the time taken to complete the experiment may be increased to some extent. Replication increases the precision of an experiment by reducing the standard deviations used to estimate factor effects. Increasing the number of replicates will decrease the error variance or mean square due to error. Replication will yield better results in the long run. Therefore it is always best to 'Do it right the first time or you'll just have to do it later!'.

8.1.10 Improve the efficiency of experimentation using blocking strategy

Blocking can be used to minimize experimental results being influenced by variations from shift-to-shift, day-to-day or machine-to-machine. The blocks can be batches of different shifts, different machines, raw materials and so on. Shainin's multi-variate charts could be a useful tool for identifying those variables which cause unwanted sources of variability. For example, a metallurgist wishes to improve the strength of a certain steel component. Four factors at 2-levels each were considered for the experiment. An eight trial experiment was chosen, but it was possible to run only four experimental trials per day. Hence each day was treated as a separate block, with the purpose of reducing day-to-day variation. It is important that the experimental trials within the block must be as homogeneous as possible.

In the context of ED, one usually has to obtain blocking generator(s) prior to applying blocking strategy. In order to obtain the blocking generators, it is advised to decide the number of blocks needed for the experiment and also the block size. It is important to ensure that the block generators are not confounded with the main effects and also two-factor interaction effects. Box et al. (1978) provide a useful table which illustrates the number of blocks, block size, recommended block generators, the number of experimental trials and the resolutions of the blocked design.

8.1.11 Understanding the confounding pattern of factor effects

The confounding pattern is often overlooked by many experimenters who use Taguchi OA designs, Plackett–Burmann designs or highly fractionated factorial designs. If we study three factors at 2-levels using four runs, the main effects will be confounded with 2-factor interactions. In other words, the estimates of main effects cannot be separated out from the interactions. It is always dangerous to run such a low resolution fractional factorial design. In the above case, we generally assign factor A to column 1, factor B to column 2 and factor C to column 3. In fact, column 3 can also be obtained due to the interaction between factors A and B. In other words, main effect C is confounded with interaction AB. If column 3 is significant from the statistical analysis, then we don't know whether the effect is the result of C, AB or both.

Confounding can be avoided by carefully choosing high resolution fractional designs, but the cost factor will go up due to the large size of the experiment. The challenge here is to find the balance between the size of the experiment and the information gained from the experiment. An understanding of confounding structures (also called alias structures) can be a tremendous asset to the experimenter.

8.1.12 Perform confirmatory runs/experiments

There is a tendency to eagerly grab the results and rush out to production and say, 'We have the answer! This will solve the problem!' Before doing that, it is important to take the time to verify the outcome of your experiment using confirmatory runs. A confirmatory run or experiment is necessary in order to verify the results of the experiment from the statistical analysis. If conclusive results have been obtained, it is then recommended to take improvement actions on the process under investigation. In contrast, if the results do not turn out as expected, further investigation would then be required. Some of the possible causes for not achieving the objective of the experiment are:

- wrong choice of ED for the experiment
- incorrect choice of quality characteristic (or response) for the experiment
- important factors which influence the response of interest were not as yet identified
- presence of non-linear or curvature effect of factors on the response of interest
- inadequate control of noise factors which cause unpleasant variation in the process under investigation
- lack of expertise for the user in the statistical analysis, etc.

8.2 Summary

Industrial experiments can be employed in all manufacturing organizations with the purpose of improving product and process quality. Both European and Western manufacturers have reported a number of successful industrial experiments. However, research has shown that very few engineers in today's industrial world are aware of industrial experiments for tackling manufacturing-process quality-control problems such as reducing scrap rate, quality costs, process variability, product development time and improving process yield, reliability, and customer satisfaction. Moreover, many engineers would not know when to utilise industrial experiments for tackling a particular quality control problem. In other words, there is a need to classify quality and engineering problems based on their potential to benefit from the use of the industrial experiments. This is an area with a lot of potential for further research.

This chapter provides some useful guidelines to engineers for making industrial experiments successful in their own organizations. The author believes that these guidelines will increase the chances for making break-through improvements in product and process quality. The key points for making your experiments successful can be therefore summarized as:

1. get a clear understanding of the problem
2. project selection
3. conduct exhaustive and detailed brainstorming session
4. teamwork and selection of a team for experimentation
5. select the continuous measurable quality characteristics for the experiment
6. choice of an appropriate ED
7. iterative experimentation
8. randomize the experimental trial order
9. replicate to dampen the effect of uncontrolled variation
10. improve the efficiency of experimentation using blocking strategy
11. understanding the confounding pattern of factor effects
12. perform confirmatory runs/experiments.

Exercises

1. Why unclear experimental objectives would lead to lost time and money?
2. What factors should be considered for the selection of an experimental design project?
3. Why brainstorming is important in the context of ED?
4. What are the advantages of choosing measurable quality characteristics over the attribute characteristics?
5. Why experiments must be conducted in an iterative manner?
6. Why do we need to perform confirmatory run/experiment?

References

Anderson, M.J. and Kraber, S.L. (1999). Eight Keys to Successful DOE, *Quality Digest*, July (downloaded from 'www.statease.com').

Anderson, M.J. (2000). 'Success with DOE', *Quality*, April, 39(4), 38–44.

Antony, J. (1996). Likes and Dislikes of Taguchi Methods, *Journal of Productivity*, 37(3), 477–481.

Antony, J. (1997). Experiments in Quality, *Journal of Manufacturing Engineer*, IEE, 76(6), 272–275.

Antony, J. (1998). Some Key Things Industrial Engineers Should Know About Experimental Design, *Logistics Information Management*, 11(6), 386–392.

Antony, J. (1999). Ten Useful and Practical Tips for Making Your Experiments Successful, *The TQM Magazine*, 11(4), 252–256.

Bhote, K.R. (1988). DOE – The High Road to Quality, *Management Review*, 27–33.

Box, G.E.P., Hunter, W.G. and Hunter, J.S. (1978). *Statistics for Experimenters*. NY, John Wiley.

Hansen, R.C. Success with Designed Experiments for Industry, *ASQ's 50th Annual Quality Congress Transactions*, pp. 718–728.

Kraber, S.L. (1998). Keys to Successful Designed Experiments, *ASQ's 52nd Annual Quality Congress Transactions*, May 4–6. Pennsylvania, pp. 119–123.

Montgomery, D.C. (1991). *Design and Analysis of Experiments*, NY, John Wiley and Sons.

Schmidt, S.R. and Launsby, R.G. (1992). *Understanding Industrial Designed Experiments*. Colorado Springs, Colorado, Air Academy Press.

Taguchi, G. (1986). *Introduction to Quality Engineering*. Tokyo, Japan, Asian Productivity Organization.

Taguchi, G. and Yokoyama, K. (1993). Taguchi Methods Design of Experiments, *Quality Engineering Series* (Vol. 4). American Supplier Institute (ASI) Press, ASI.

Verseput, R. (1998). DOE Requires Careful Planning, *R&D Magazine*, pp. 71–72.

9

Case studies

9.1 Introduction

This chapter presents a collection of real industrial case studies. The case studies illustrated in the chapter are well-planned experiments and not simply a few experimental trials to explore the effects of varying one or more factors at a time. The case studies will provide a good foundation for students, researchers and practitioners on how to go about carrying out an experiment in real industrial settings. The case studies will cover the nature of the problem or objective of the experiment, list of factors, their levels, response of interest, choice of a particular design (i.e. number of trials used), analysis using Minitab software, interpretation of results and benefits gained from the experiment. These case studies will increase the awareness of the application of experimental design techniques in industries and its potential in tackling process optimization and variability problems.

9.2 Case studies

9.2.1 Optimization of a radiographic quality welding of cast iron

Objective of the experiment

The objective of the experiment was to identify the significant welding parameters and determine the optimal parameter settings which gave minimum crack length.

Selection of the response function

The response of interest for the experiment was crack length measured in centimetres.

List of factors and interactions of interest for the experiment

Five main effects and 2 two-order interactions were identified from a thorough brainstorming session. The list of main and interaction effects are shown below:

Main effects. Current (A), Bead length (B), Electrode make (C), V-groove angle (D) and Welding method (E)
Interaction effects. A × B and B × C

Levels of parameters and their ranges

Each parameter was studied at 2-levels. The ranges of welding parameters are shown in Table 9.1.

Choice of design and number of experimental trials

As the number of factors is more than four, it was decided to select a fractional factorial design rather than a full factorial design. The number of degrees of freedom for studying both main effects and interactions is equal to seven. The closest number of experimental trials that can be employed for this study is eight. This means it is a $2^{(5-2)}$ fractional factorial design in which main effects are confounded with two-factor interactions. In other words, the design resolution of this design is III.

Design generators and the confounding structure of the design

Design generators

$$D = AC \quad \text{and} \quad E = ABC$$

Defining relationship

$$I = ACD \times I = ABCE \quad \text{and} \quad I = BDE$$

Table 9.1 List of factors and their ranges for the experiment

Welding parameters	Labels	Low level	High level
Current	A	110	135
Bead length	B	20	30
Electrode make	C	X	Y
V-groove angle	D	45	60
Welding method	E	1	2

Confounding pattern

$$A = CD = BCE$$
$$B = ACE = DE$$
$$C = AD = ABE$$
$$D = AC = BE$$
$$E = ABC = BD$$
$$AB = CE = ADE = BCD$$
$$AC = BE, \quad BC = AE = ABD = CDE$$

Uncoded design matrix with response values

The uncoded design matrix showing all the real factor settings along with the respective response values is shown in Table 9.2. Each trial condition was replicated twice to create adequate degrees of freedom for the error term. Randomization strategy was employed to minimize the effect of lurking variables and undesirable external influences induced into the experiment. As we can see from Table 9.2, welding parameter C (electrode make) was assigned to column 1 as it was not practical to change the levels of this factor frequently.

Analysis and Interpretation of results

The first step was to check the data for normality assumptions. This was achieved by constructing normal probability plot (NPP) of residuals (Figure 9.1). The plot suggests that the data follow a normal distribution. The analysis part involves the determination of significant main and interaction effects, followed by the selection of optimal welding parameter settings which yields minimum crack length. In order to identify the most important main and interaction effects, it was decided to use a Pareto plot of effects (Figure 9.2).

Table 9.2 Uncoded design matrix with response values

Standard order	C	B	A	D = AC	E = ABC	Crack length (cm)	
1 (5)	X	20	110	60	1	9	12
2 (3)	Y	20	110	45	2	7	8
3 (8)	X	30	110	60	2	7	5
4 (2)	Y	30	110	45	1	13.5	12.0
5 (6)	X	20	135	45	2	10	9
6 (1)	Y	20	135	60	1	6.5	8
7 (7)	X	20	135	45	1	7	6
8 (4)	Y	20	135	60	2	7.5	8

Note: ( ) represents the order in which the experimental runs werye carried out.

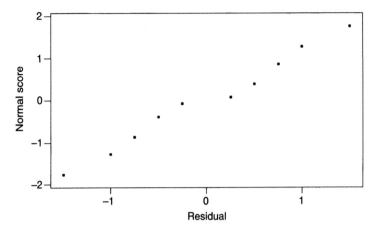

Figure 9.1 Normal probability plot of residuals.

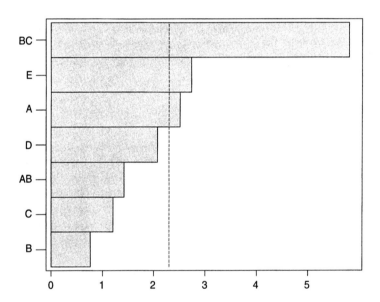

Figure 9.2 Pareto plot of effects from the experiment.

Figure 9.2 indicates that main effects A and E and interaction effect BC were considered to be real (or active). In order to analyse interaction between B and C, it was decided to use an interaction plot shown in Figure 9.3.

Figure 9.3 indicates that there is a strong interaction between B and C. Moreover, it can be observed from the figure that crack length is minimum when B is kept at high level setting and C at low level setting. In order to determine the optimal welding parameter settings which yields minimum

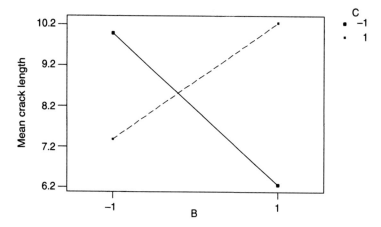

Figure 9.3 Interaction plot of B vs C.

crack length, a main effects plot is constructed (Figure 9.4). The optimal settings for minimizing crack length is:

A: +1 (high level)
B: +1 (high level)
C: −1 (low level)
D: +1 (high level)
E: +1 (high level)

Confirmatory trials

Three confirmatory trials based on the optimal settings were performed and crack lengths of 0.31 mm, 0.46 mm and 0.32 mm were observed. The results of the study have demonstrated a significant improvement to the process and a significant reduction in scrap and rework was achieved.

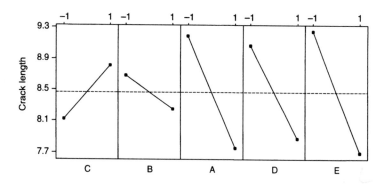

Figure 9.4 Main effects plot for crack length.

9.2.2 Reducing process variability using Experimental Design technique objective of the experiment

The objective of the experiment was to identify the most important process parameters that affect variability in response.

Selection of the response

The response of interest for the experiment was expulsion force measured in kilograms (kg). Here expulsion force is the force required to expel the device or component from a certain tube.

List of process parameters and their levels

Seven process parameters were identified from a brainstorming session with people from production, maintenance, quality, design and shop-floor. As part of initial investigation of the study, it was decided to study the effect of main effects on variability in expulsion force. The list of parameters used for the experiment and their levels are illustrated in Table 9.3.

Choice of design and number of experimental trials required for the experiment

For this study, seven factors were thought to have some impact on variability in expulsion force. A full factorial experiment would require a total of 128 experimental trials. Owing to limited budget and the top management needing a speedy response to this investigation, it was decided to use a highly fractionated factorial design. Here the objective was to identify the key process parameters so that further smaller experiments could be carried out to study the interactions among the key parameters. The number of degrees of freedom associated with seven factors at 2-levels is seven. Hence the number of degrees of freedom required for the experiment must be greater than seven. The closest number of experimental trials that can be employed for this study is eight, i.e. a $2^{(7-4)}$ fractional factorial design was selected.

Table 9.3 List of process parameters and their levels

Process parameters	Labels	Low level	High level
Position of the cam	A	Forward (F)	Backward (B)
Drum temperature	B	84	104
Time	C	68	72
Type of material	D	1	2
Clearance	E	0.006	0.012
Machine alignment	F	134	130
Header temperature	G	190	210

Design generators and resolution

$C = -AB$
$E = -AD$
$F = -BD$
$G = -ABC$

As the main effects are confounded with two-factor interactions, the resolution of this design is III.

Coded and uncoded design matrix with response values

The uncoded and coded design matrix with response values is shown in Table 9.4 and Table 9.5 respectively. Each trial condition was repeated five times to analyse variability.

Analysis and interpretation of results

As the objective of the experiment is to reduce variability in expulsion force, the first step is to identify which of the seven factors have impact on

Table 9.4 Uncoded design matrix with response values

Run	A	B	C	D	E	F	G	Expulsion force (kg)
1	F	84	68	1	0.006	134	190	0.990, 1.037, 0.965, 0.860, 1.086
2	B	84	72	1	0.012	134	210	0.875, 0.748, 0.959, 0.600, 0.807
3	F	104	72	1	0.006	130	210	0.924, 0.881, 0.733, 0.767, 0.873
4	B	104	68	1	0.012	130	190	0.760, 0.620, 0.669, 0.632, 0.605
5	F	84	68	2	0.012	130	210	0.741, 0.455, 0.549, 0.468, 0.646
6	B	84	72	2	0.006	130	190	0.787, 1.061, 0.607, 1.168, 0.878
7	F	104	72	2	0.012	134	190	0.508, 0.446, 0.351, 0.419, 0.421
8	B	104	68	2	0.006	134	210	0.691, 0.771, 0.940, 0.743, 0.675

Table 9.5 Coded design matrix with response values

Run	A	B	C	D	E	F	G	Expulsion force (kg)
1	-1	-1	-1	-1	-1	-1	-1	0.990, 1.037, 0.965, 0.860, 1.086
2	1	-1	1	-1	1	-1	1	0.875, 0.748, 0.959, 0.600, 0.807
3	-1	1	1	-1	-1	1	1	0.924, 0.881, 0.733, 0.767, 0.873
4	1	1	-1	-1	1	1	-1	0.760, 0.620, 0.669, 0.632, 0.605
5	-1	-1	-1	1	1	1	1	0.741, 0.455, 0.549, 0.468, 0.646
6	1	-1	1	1	-1	1	-1	0.787, 1.061, 0.607, 1.168, 0.878
7	-1	1	1	1	1	-1	-1	0.508, 0.446, 0.351, 0.419, 0.421
8	1	1	-1	1	-1	-1	1	0.691, 0.771, 0.940, 0.743, 0.675

Table 9.6 Standard deviation and ln(SD) values

Run	A	B	C	D	E	F	G	SD	ln(SD)
1	−1	−1	−1	−1	−1	−1	−1	0.085	−2.465
2	1	−1	1	−1	1	−1	1	0.136	−1.995
3	−1	1	1	−1	−1	1	1	0.081	−2.513
4	1	1	−1	−1	1	1	−1	0.0621	−2.779
5	−1	−1	−1	1	1	1	1	0.122	−2.104
6	1	−1	1	1	−1	1	−1	0.222	−1.505
7	−1	1	1	1	1	−1	−1	0.057	−2.865
8	1	1	−1	1	−1	−1	1	0.106	−2.244

variability. In order to analyse variability, both SD and ln(SD) (natural logarithms of standard deviation) were computed at each experimental design point. The results are shown in Table 9.6.

A NPP of residuals was constructed for the validity of normality assumptions (Figure 9.5). The figure shows that the data come from a normal population. Having checked the data for normality, the next step was to identify the factors which influence variability in expulsion force. Both main effects plot and Pareto plot are used to identify the key process parameters or factors which have impact on variability. The graphs (Figures 9.6 and 9.7) indicate that factor B has a significant impact on variation. In order to obtain adequate degrees of freedom for the error variance term, pooling strategy was utilized. The rule of thumb is to pool the effects with low magnitude till the error degrees of freedom is nearly half the total degrees of freedom. It was interesting to note that variability is minimum when factor B is kept at high level (Figure 9.7).

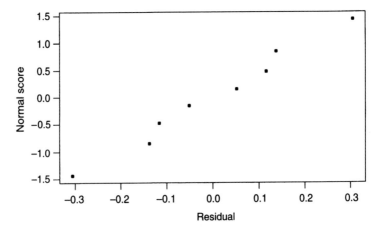

Figure 9.5 Normal probability plot of residuals for ln(SD).

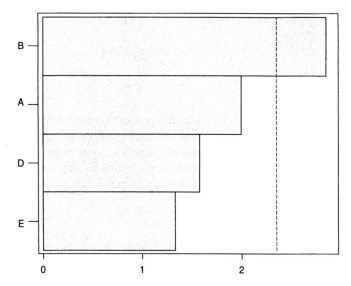

Figure 9.6 Pareto plot of effects for ln(SD).

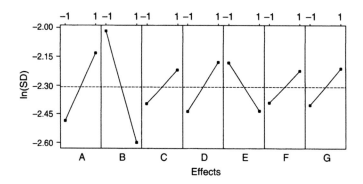

Figure 9.7 Main effects plot for ln(SD).

Determination of optimal settings to minimize variability

In order to determine the optimal settings to minimize variability, the first step was to rank the factors (in descending order of importance) which influence variability in expulsion force.

Factor B – Rank 1
Factor A – Rank 2
Factor D – Rank 3
Factor E – Rank 4
Factor G – Rank 5
Factor C – Rank 6
Factor F – Rank 7

The optimal condition based on the main effects plot was obtained as follows:

$$B_{(1)} \; A_{(-1)} \; D_{(1)} \; E_{(1)} \; G_{(-1)} \; C_{(-1)} \; F_{(-1)}$$

Confirmation trials

Fifteen samples were produced under the optimal conditions and compared against the samples produced under standard production conditions. The sample standard deviation at the optimal settings was reduced to 0.042 kg as opposed to 0.125 kg under normal production conditions. The reduction in SD was therefore estimated to be approximately 66 per cent.

Significance of the work

Due to the significant reduction in process variability, the actual capability of the process has increased from 0.86 to over 1.78. This clearly demonstrates a dramatic improvement in the process performance and thereby more reliable and consistent products can be produced by determining the optimal condition of the process under study. The benefits from this study include increased customer satisfaction, reduced warranty costs, reduced customer complaints, reduced scrap and rework, improved market share, improved process control and so on and so forth. The engineering team including production personnel, quality engineers and managers of the company are now well aware of the benefits that can be gained from the application of experimental design methods. Moreover, the awareness that has been established within the organization has built confidence among the engineers, managers and front-line workers in other areas facing similar difficulties.

9.2.3 Slashing scrap rate using fractional factorial experiments

Nature of the problem

This case study describes the application of a highly fractionated factorial design to a manufacturing process that makes electromagnetic clutch coils. The coils were made of about 0.75 mm diameter copper wire. When winding the coil to form into a solenoid, the wire is heated to around 180 °C, which turns the insulation into an adhesive that bonds the wires together. However, the company that produces these coils was facing a quality problem in the form of high scrap rate, rework, etc. which incurred huge failure costs to the company. Hence it was important for the company to find out what was causing this.

Objective of the experiment

The objective of the experiment was to identify the most important machine parameters which gave minimum scrap rate (per cent).

Selection of the response

The response of interest for the experiment was the percentage of rejects.

List of process parameters and their levels

With limited budget and resources, it was important to study the effect of seven parameters on the percentage of rejects. To minimize the number of experimental trials, each factor was studied at 2-levels: low and high. The list of process (or machine) parameters and their levels are shown in Table 9.7.

Coded design matrix with response values for the experiment

The coded design matrix describes all the process parameter combinations at their respective levels and the order in which the runs or experimental trials were performed. A total of 2500 samples were used for each trial condition, and the percentage of rejects recorded for the analysis. In order to minimize the effect of lurking variables, randomization strategy was employed. The results of the experiment are shown in Table 9.8.

Table 9.7 List of parameters and their levels used for the experiment

Process parameters	Labels	Low level	High level
Felt lubrication	A	Dry	Soaked
Wire diameter	B	0.75 mm	0.76 mm
Friction on pulley	C	Low	High
Brake tension	D	1.5 kg	2 kg
Winding width	E	High	Low
Dirt buildup	F	Unclean	Clean
Axial start position	G	A	B

Table 9.8 Experimental layout with response values

Standard order	A	B	C	D	E	F	G	Rejects (%)
1	−1	−1	−1	−1	−1	−1	−1	1.08
2	1	−1	1	−1	1	−1	1	2.52
3	−1	1	1	−1	−1	1	1	1.12
4	1	1	−1	−1	1	1	−1	1.20
5	−1	−1	−1	1	1	1	1	3.04
6	1	−1	1	1	−1	1	−1	2.76
7	−1	1	1	1	1	−1	−1	1.00
8	1	1	−1	1	−1	−1	1	1.92

Analysis and interpretation of results

The analysis part involves the identification of the most important machine (or process) parameters which causes the problem. In order to identify the key parameters, a Pareto plot was used (Figure 9.8).

The above graph shows that machine parameters B, D and G are statistically significant at 10 per cent significance level. Machine parameters A, C, E and F have a relatively trivial effect. Having identified the key parameters, the next step was to determine the best settings which yields the best performance. For the present study, a main effects plot was constructed (Figure 9.9). The graph clearly shows that the optimal levels of all the parameters except B (the most important) is -1 (low level setting). The optimal settings for the parameters were obtained as:

$$A_{(-1)} \ B_{(1)} \ C_{(-1)} \ D_{(-1)} \ E_{(-1)} \ F_{(-1)} \ G_{(-1)}$$

Confirmation runs

For confirmation runs, five batches of 500 samples were used. The results of the confirmation runs were remarkable due to a very significant reduction in the scrap rate of only 0.37 per cent. As a result of this significant reduction in scrap, the company expects to save more than $120,000 per annum. Moreover, the quality and production personnel of the organization have been persuaded to extend the application of simple experimental design methods to other core processes.

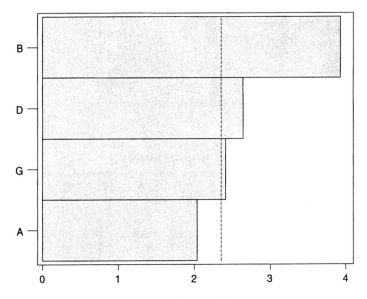

Figure 9.8 Pareto plot of effects for the experiment.

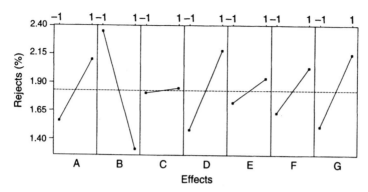

Figure 9.9 Main effects plot for the experiment.

9.2.4 Optimizing the time of flight of a paper helicopter

Objective of the experiment

The objective of the experiment was to determine the optimal settings of the design parameters which would maximize the time of flight of a paper helicopter.

Description of the experiment

The experiment was carried out by the author in a class room for a post-graduate course in quality management with the aim of demonstrating how Design of Experiments can be employed for optimizing the design parameters of a simple paper helicopter. The experiment requires paper, scissors, ruler, paper clip, measuring tape and a stopwatch. It would take approximately 5 to 6 hours to design, conduct and analyse the results of the experiment. The model of a paper helicopter design is shown in Figure 9.10.

Selection of the response

The response of interest to the experimenter in this case was the time of flight measured in seconds.

List of design parameters and their levels

Six design parameters were chosen for this experiment. In order to make the experiment simple, it was decided to study each design parameter at 2-levels. Design parameters at 3-levels are more complicated to teach in the first place and moreover the author strongly believes that it might turn off engineers from learning Design of Experiments any further. The logic behind a simple but practical experiment of this nature is to demonstrate the importance of experimental design and to illustrate how it works in real life situations. Table 9.9 presents the list of design parameters and their levels selected for the experiment.

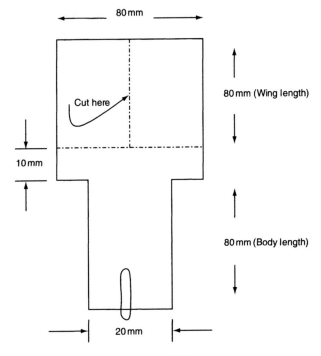

80 mm

Cut here

80 mm (Wing length)

10 mm

80 mm (Body length)

20 mm

Figure 9.10 Model of a paper helicopter design.

Table 9.9 List of design parameters and their levels

Design parameters	Labels	Low level (−1)	High level (+1)
Paper type	A	Normal	Bond
Body length	B	80 mm	130 mm
Wing length	C	80 mm	130 mm
Body width	D	20 mm	35 mm
Number of clips	E	1	2
Wing shape	F	Flat	Angled 45° up

Apart from the main effects, three interaction effects were also of interest to analyse for the experiment. These are: (1) B × C, (2) B × D, and (3) A × E.

In order to minimize the effect of noise parameters such as draft and operator on the time of flight, extra caution was taken during the experiment. The experiment was conducted in a closed room to dampen the effect of draft. The same operator was responsible to minimize the reaction time of hitting the stopwatch when the helicopter is released and when it hits the floor.

Choice of design and design matrix for the experiment

As we are interested in studying six main effects and three interaction effects, the total degrees of freedom is equal to nine. The closest number of

Table 9.10 Uncoded design matrix with response values

Run	A	B	C	D	E	F	Time of flight (sec)
1 (6)	Normal	80	80	20	1	Flat	2.49
2 (9)	Bond	80	80	20	2	Flat	1.80
3 (11)	Normal	130	80	20	2	Angled	1.82
4 (15)	Bond	130	80	20	1	Angled	1.99
5 (12)	Normal	80	130	20	2	Angled	2.11
6 (2)	Bond	80	130	20	1	Angled	1.96
7 (16)	Normal	130	130	20	1	Flat	3.19
8 (14)	Bond	130	130	20	2	Flat	2.27
9 (10)	Normal	80	80	35	1	Angled	2.12
10 (1)	Bond	80	80	35	2	Angled	1.58
11 (7)	Normal	130	80	35	2	Flat	2.15
12 (3)	Bond	130	80	35	1	Flat	2.05
13 (8)	Normal	80	130	35	2	Flat	2.60
14 (4)	Bond	80	130	35	1	Flat	2.09
15 (5)	Normal	130	130	35	1	Angled	2.63
16 (13)	Bond	130	130	35	2	Angled	2.18

experimental trials that can be employed for the experiment is 16 (i.e. $2^{(6-2)}$ fractional factorial design). This means that only a quarter replicate of a full factorial experiment is needed for the study. The uncoded design matrix for the experiment along with recorded response values corresponding to each trial condition is shown in Table 9.10.

Statistical analysis and interpretation of results

Prior to carrying out any statistical analysis, the first step was to check the data for normality assumptions. A NPP of residuals was constructed (Figure 9.11) which indicates that the data come from a normal population. The next stage of

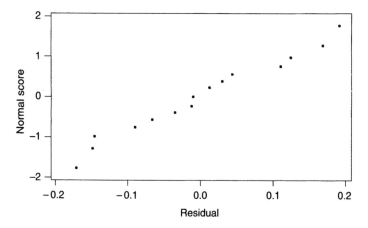

Figure 9.11 Normal probability plot of residuals.

the analysis was to identify which of the main or/and interaction effects have significant impact on the time of flight. It was decided to use a Pareto plot using Minitab software. Minitab plots the effects in decreasing order of the absolute value of the standardized effects and draws a reference line on the chart. Any effect that extends the reference line appears to be statistically significant. The Pareto plot of the effects (Figure 9.12) shows that the main effects (A, C, F and E) are statistically significant (assume 5 per cent significance level).

None of the interactions appear to be statistically significant. The interaction between B and C was not statistically significant at 5 per cent significance level, though it appeared to be important in the interaction graph (Figure 9.13).

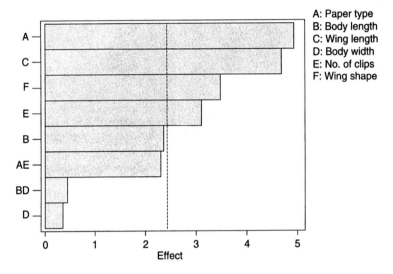

Figure 9.12 Pareto plot of the effects from the experiment.

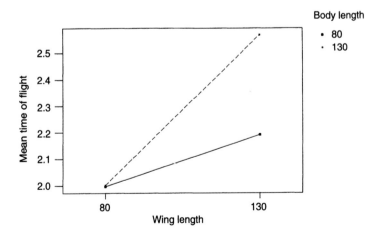

Figure 9.13 Interaction plot between wing length and body length.

It was rather interesting to observe that body width has no influence on the time of flight.

Determination of optimal design parameters

Having identified the significant design parameters which influence the time of flight, the next step is to determine the optimal settings that maximizes the time of flight. As none of the interaction effects were statistically significant, the best levels of each parameter can be readily obtained from a main effects plot (Figure 9.14). The final optimal settings of the design parameters is:

Design parameter A – Low Level (normal paper)
Design parameter B – High level (130 mm)
Design parameter C – High level (130 mm)
Design parameter D – Low level (20 mm)
Design parameter E – Low level (no. of clips = 1)
Design parameter F – Low level (flat)

It is quite interesting to note that the time of flight was maximum when wing length and body length were kept at high levels.

Predicted model for time of flight

A simple regression model is developed based on the significant effects. It is important to note that the regression coefficients in the model are half the estimates of the effects. The regression model for the time of flight can be therefore written as:

$$\hat{y} = \beta_0 + \beta_1(A) + \beta_2(C) + \beta_3(F) + \beta_4(F) \tag{9.1}$$

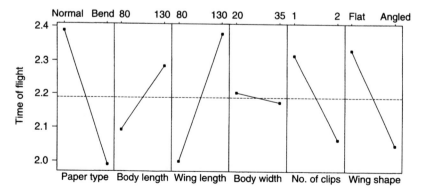

Figure 9.14 Main effects plot of the design parameters.

where β_0 = overall mean time of flight = 2.19, β_1 = regression coefficient of factor A (paper type), β_2 = regression coefficient of factor C (wing length), β_3 = regression coefficient of factor F (wing shape) and β_4 = regression coefficient of factor E (no. of clips).

The predicted model for time of flight is therefore given by

$$\hat{y} = 2.19 + (-0.20 \times -1) + (0.19 \times 1) + (-0.14 \times -1) + (-0.13 \times -1)$$

$$\hat{y} = 2.85 \sec$$

Confirmatory runs

A confirmatory experiment was carried out to verify the results from the analysis. Ten helicopters were made based on the optimal settings of the design parameters. The average flight time was estimated to be 3.09 sec with a standard deviation of 0.35 sec.

Confidence Interval (based on 95 per cent confidence level) $= \bar{y} \pm 3 \times \frac{SD}{\sqrt{n}}$, where 'SD' is the sample standard deviation, $\bar{y}$ is the sample mean and n is the sample size.

Therefore,

$$\text{Confidence interval} = 3.09 \pm 3 \times 0.11$$

$$= 3.09 \pm 0.33$$

$$= (2.76, 3.42)$$

As the predicted value (2.85 sec) for the optimal settings falls within the above confidence interval, we can conclude that the predicted model is sound.

Significance of the work

The purpose of this case study is to demonstrate the importance of teaching experimental design methods to people with limited skills in statistics for tackling variability and poor process performance problems. This experiment is quite old in its nature and has already been widely used for some time by many statisticians for teaching purposes. Nevertheless the focus here was to minimize the statistical jargon associated with the technique and bring modern graphical tools for better and rapid understanding of the results to non-statisticians. The students of the class found this experiment very interesting specifically in terms of selecting the design, conducting the experiment and interpreting the results. Many students were quite astounded with the use of simple but powerful graphical tools and its reduced involvement of number crunching.

9.2.5 Optimizing a wire bonding process using Design of Experiments

Objective of the experiment

The following are the objectives of the experiment:

- to determine the optimal process parameter settings for enhanced strength
- to develop a mathematical model which relates the wire pull strength and the key process parameters which influence the strength.

Description of the experiment

This case study illustrates a wire bonding process making a physical connection between the die and the lead. The purpose of this study was to increase the wire pull strength due to an increased number of customer complaints on broken wires.

Selection of the response

The response of interest to the experimenter was wire pull strength expressed in grams.

Identification of process variables for experimentation

The following process variables were identified from a thorough brainstorming session. People from quality department, production department and operators were involved in the session. Each process variable was studied at 2-levels as part of initial investigation. Table 9.11 presents the list of parameters used for the experiment.

The following interactions were of interest to the experimenter:

1. B × C, 2. A × C, 3. A × D and 4. A × B.

All three-order and higher-order interactions are neglected.

Table 9.11 List of process parameters used for the experiment

Process variables	Labels	Low level	High level	Unit
Power	A	100	150	mW
Temperature	B	140	200	°C
Bonding time	C	15	25	Ms
Bonding force	D	3	9	grams

Choice of design and experimental layout

The choice of design is dependent on the number of main and interaction effects to be studied, cost and time constraints, required design resolution, etc. As the total degrees of freedom required for studying the four main effects and four interaction effects is equal to eight, the most suitable design for this experiment was a 2^4 full factorial experiment. This allows one to estimate all the main effects and interactions independently. Each trial condition was randomized to minimize the effect of lurking variables. The uncoded design matrix along with response values is shown in Table 9.12. The next step illustrates how the results of the experiment have been analysed.

Statistical analysis and interpretation

In order to identify the significant main effects and interaction effects, it was decided to use a NPP of effects. Those effects which fall off the straight line are deemed to be statistically significant and those which fall along the straight line are deemed to be statistically insignificant. The NPP of effects is shown in Figure 9.15. The figure shows that main effects A, B, D and interaction effect AD are statistically significant at 5 per cent significance level. In order to determine the best levels for A and D, it was important to analyse the interaction effect (A × D). Figure 9.16 illustrates the interaction plot between A and D.

The non-parallel lines indicate that there is a strong interaction between the process variables A and D. As we can observe from the plot, the effect of bonding force on the pull strength is different at low and high levels of power. Minimum variability in pull strength is observed at high level of power. On

Table 9.12 Uncoded design matrix for the experiment

Trial no.	A	B	C	D	Pull strength
1 (7)	−1	−1	−1	−1	7.4
2 (11)	1	−1	−1	−1	6.5
3 (5)	−1	1	−1	−1	8.2
4 (15)	1	1	−1	−1	8.8
5 (2)	−1	−1	1	−1	7.6
6 (9)	1	−1	1	−1	6.8
7 (10)	−1	1	1	−1	8.4
8 (16)	1	1	1	−1	8.6
9 (3)	−1	−1	−1	1	9.4
10 (13)	1	−1	−1	1	8.0
11 (4)	−1	1	−1	1	9.8
12 (1)	1	1	−1	1	8.9
13 (6)	−1	−1	1	1	9.0
14 (12)	−1	−1	1	1	7.9
15 (8)	1	1	1	1	10.1
16 (14)	−1	1	1	1	9.1

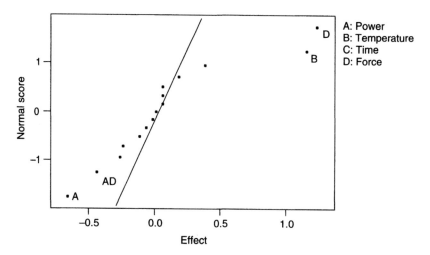

Figure 9.15 Normal probability plot of effects.

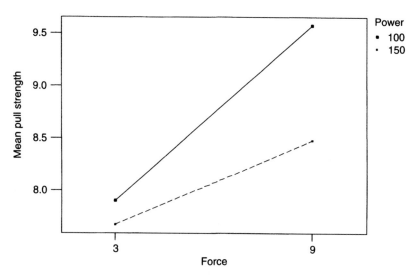

Figure 9.16 Interaction between power (A) and force (D).

the other hand, mean strength is higher at high level of bonding force (9 gms) and low level of power (100 mW).

In order to identify the optimal settings of process parameters which gives maximum pull strength, a main effects plot was constructed (Figure 9.17).

Table 9.13 presents the optimal settings of bonding process parameters that would yield maximum strength. It is important to note that bonding time has no influence whatsoever on the pull strength. Hence it was decided to select

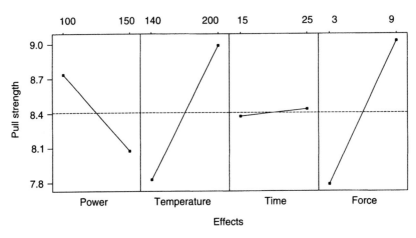

Figure 9.17 Main effects plot of wire bonding experiment.

Table 9.13 Optimal condition of the wire bonding process

Process parameters	Uncoded level	Coded level
Power	100 mW	−1
Temperature	200 °C	1
Bonding time	15 ms	−1
Bonding force	9 g	1

15 ms as optimal value compared to 25 ms. Here bonding time can be treated as a cost adjustment factor.

Model development based on the significant factor/interaction effects

Having identified the significant main and interaction effects which influence the pull strength, it was considered important to develop a simple regression model which provides the relationship between the pull strength and the critical effects. The use of this model is to predict the pull strength for different combinations of wire bonding process parameters at their best levels. It is important to note that for process parameters at 2-levels, the regression coefficients are half the estimates of the effects. Table 9.14 presents the estimates of significant effects and regression coefficients. The regression model for the wire bonding process as a function of significant main and interaction effects is given by:

$$\hat{y} = \beta_0 + \beta_1(A) + \beta_2(B) + \beta_4(D) + \beta_{14}(A \times D)$$
$$\hat{y} = 8.41 - 0.33A + 0.58B + 0.62D - 0.22AD$$

Table 9.14 Estimates of effects and regression coefficients

Process parameters/interactions	Estimate of effects	Regression coefficients
A	−0.663	−0.33
B	1.162	0.58
D	1.237	0.62
AD	−0.438	−0.22

where $\hat{y}$ is the predicted pull strength.

The predicted pull strength based on the significant factor and interaction effects (based on the optimal condition) is hence given by:

$$\hat{y} = 8.41 - 0.33(-1) + 0.58(1) + 0.62(1) - 0.22(-1)(1)$$
$$\hat{y} = 10.16$$

Confirmation trials at the optimal condition have yielded a mean pull strength of 10.25 g. Ninety five per cent confidence interval of the mean pull strength is given by:

$$95 \text{ per cent CI} = \bar{y} \pm 3 \text{ (s.e.)}, \quad \text{where s.e. is the standard error}$$
$$= 10.25 \pm 3(0.19)$$
$$= (9.68, \ 10.82)$$

As the predicted value falls within this interval, it is fair to conclude that the predicted model for pull strength is sound and practical.

Conclusion

This case study presents a study performed on a certain wire bonding process using DOE with two objectives in mind. The first objective of the experiment was to understand the process by identifying the key wire bonding process parameters and the interactions of interest. The second objective was to develop a regression model for predicting the pull strength at the optimal condition of the process. The results of the study have shown an improvement in pull strength by over 20 per cent over the existing production conditions.

9.2.6 Training for Design of Experiments using a catapult

The purpose of this case study was to provide an insight into the process of understanding the role of DOE as part of a training program to a group of engineers and managers in a world class company. The results of the experiment have been extracted from a simple full factorial experiment performed using a catapult. The results of the experiment were analysed using Minitab software for rapid and easier understanding of the results.

Objective of the experiment

The objective of the experiment was to maximize the in-flight distance.

Selection of response

The response of interest to the team was in-flight distance measured in meters.

List of factors and their levels used for the experiment

Four factors (stop position, peg height, release angle and hook position) were studied at 2-levels. These factors were identified from a brainstorming session facilitated by the author. The levels for factors such as type of ball, type of rubber band and cup position were kept constant. This implies that pink ball, 6th cup position and brown rubber band were used throughout the experiment. Table 9.15 presents the list of factors and their levels used for the experiment.

Choice of design and experimental layout for the experiment

It was decided to perform a full factorial experiment so that it allows us to study all the main and interaction effects. The experiment was replicated twice to capture the variation due to experimental set up and airflow in the room. Each trial condition was randomized to minimize the bias induced into the experiment. The results of the experiment along with response values are shown in Table 9.16.

Having performed the experiment, the next step was to analyse and interpret the results so that necessary actions could be taken accordingly. The analysis of the experiment is often dependent on its objective. In this case, the objective was to identify the factors which affect the in-flight distance. The team used Minitab to analyse the data from the experiment. This is the focus of the next section.

Statistical analysis and interpretation of results

Prior to carrying out the statistical analysis, the first step was to check the data for normality assumptions. A NPP of residuals (Figure 9.18) was constructed using Minitab software. It can be seen in Figure 9.18 that all the points on the

Table 9.15 List of factors and their levels for catapult experiment

Factors	Labels	Low level	High level
Release angle	RA	180	Full
Peg height	PH	3	4
Stop position	SP	3	5
Hook position	HP	3	5

Table 9.16 Results of the full factorial experiment

Trial no.	RA	PH	SP	HP	Distance (m)
1 (4)	−1	−1	−1	−1	3.62, 3.64
2 (8)	1	−1	−1	−1	4.01, 4.06
3 (11)	−1	1	−1	−1	4.16, 4.60
4 (7)	1	1	−1	−1	4.70, 4.90
5 (1)	−1	−1	1	−1	3.80, 3.83
6 (10)	1	−1	1	−1	4.37, 4.40
7 (3)	−1	1	1	−1	4.74, 4.77
8 (15)	1	1	1	−1	5.32, 5.58
9 (2)	−1	−1	−1	1	4.26, 4.13
10 (14)	1	−1	−1	1	4.74, 4.94
11 (6)	−1	1	−1	1	4.80, 5.02
12 (13)	1	1	−1	1	5.20, 5.55
13 (16)	−1	−1	1	1	4.46, 4.67
14 (5)	1	−1	1	1	5.12, 5.50
15 (12)	−1	1	1	1	4.80, 4.85
16 (9)	1	1	1	1	5.80, 5.91

Note: ( ) represents the experimental trials/runs in random order.

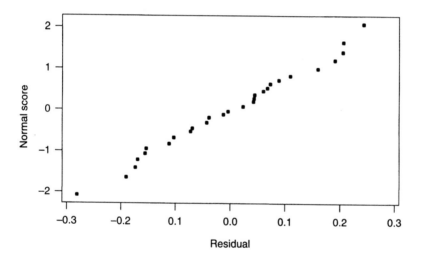

Figure 9.18 Normal probability of residuals.

normal plot come close to forming a straight line. This implies that the data are fairly normal. The next step is to identify the most significant main and interaction effects which influence the distance.

In order to identify the most important effects, it was decided to use a Pareto plot. The Pareto plot (Figure 9.19) shows that all the main effects (RA, PH, HP and SP) and one interaction effect (PH × HP) are deemed to be active. In order to interpret the interaction between PH and HP effectively, an interaction plot was constructed (Figure 9.20).

The interaction plot indicates that the effect of hook position (HP) at different levels of peg height (PH) is not the same. This implies that there is

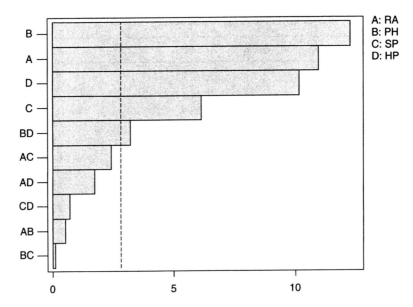

Figure 9.19 Pareto plot of effects from a catapult experiment.

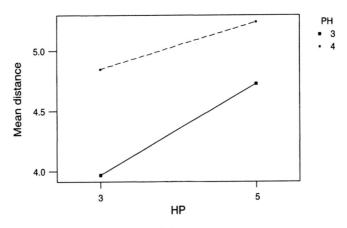

Figure 9.20 Interaction plot – HP × PH.

a strong interaction between these two factors. The graph also shows that maximum distance was achieved when HP is kept at position 5 and PH at position 4.

Determination of optimal factor settings

In order to arrive at the optimal condition, the mean distance at each level of the control factor was analysed. A main effects plot was constructed to

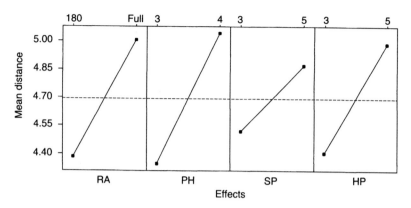

Figure 9.21 Main effects plot for the catapult experiment.

identify the best levels of the factors (Figure 9.21). The best settings of the factors for maximizing the in-flight distance is (Figure 9.21):

Release angle – Full
Peg height – Position 4
Stop position – Position 5
Hook position – Position 5

It is worthwhile noting that the optimal condition is one which corresponds to trial condition 16 (Table 9.16). This is due to the fact that it is a full factorial experiment, which shows all the possible combinations. It is not necessary that this is the case in many industrial experiments due to various constraints (time, cost, objective of the experiment, degree of resolution required, etc.).

Confirmatory experiment

A confirmatory experiment was carried out to verify the results from the analysis. Five observations were made at the optimal condition. The average in-flight distance was estimated to be 5.84 m. It was also observed that a change of stop position from 5 to 4 has yielded even better average results in distance (i.e. 5.96 m).

Significance of the work

The purpose of this case study was to bring the importance of teaching DOE to a group of engineers and managers in a world class organization using simple but powerful graphical tools. The focus of this study was to minimize the statistical jargon associated with DOE and to bring modern graphical tools for rapid decision-making process. The results of this experiment have provided a greater stimulus for the wider application of DOE by engineers within this organization in other core processes for tackling variability related and process optimization problems.

9.2.7 Optimization of core tube life using designed experiments

This case study presents two different experiments – the first experiment was performed by the engineering team within the company and the second one was performed by the author with the help of operations personnel within the company. The product of concern in this case study is core tube used within a solenoid-operated directional control valve. The problem with this product was that the life was short when subjected to hydraulic fatigue test. The core tube assembly is welded and then machined prior to final assembly of the system. The company uses laser welding for core tube assembly and therefore most of the factors affecting the life of these core tubes were related to the laser welding process. Laser welding was chosen for the core tube assembly because the technique affords a high degree of repeatability, predictability and good control of penetration depth.

Company's first attempt to experimental approach

The first experiment was performed by the engineering team consisting of quality engineer, design engineer, production engineer and operator. In order to keep the experimental budget minimum, it was decided to study all factors (or process parameters) at 2-levels. Three process parameters were chosen by the team which they believed to have some impact on the life of the core tube. The response of interest to the team was the fatigue life of the core tube, expressed in number of cycles (in millions).

The team has decided to study only the effects of three laser welding process parameters. Interactions among the parameters were of interest to the team. A $2^{(3-1)}$ fractional factorial design was chosen for the experiment. Table 9.17 illustrates the list of welding process parameters used for the experiment.

Table 9.18 shows the experimental layout for the optimization of core tube life. The experimental layout displays the number of experimental trials, process parameters and the response values corresponding to each experimental design point.

Table 9.17 Process parameters for the experiment

Process parameter	Label	Low level	High level	Units
Weld speed	A	1.5	2.0	Rev/sec
Ramp out	B	1	2	Seconds
Ramp in	C	0.5	1.5	Seconds

Table 9.18 Experimental layout for the experiment

Run	A	B	C	No. of cycles (in million)
1	1.5	1	1.5	1.92
2	2.0	1	0.5	4.80
3	1.5	2	0.5	2.24
4	2.0	2	1.5	6.93

The desired number of cycles on average is about 8.5. This is to conform with the requirements of the National Fluid Power Association Standards. None of the above trial conditions yielded a value more than seven million cycles. The analysis of results indicates that weld speed has the highest impact on core tube life and ramp in has the least influence. Table 9.19 presents the effects of the laser welding process parameters.

The objective of the experiment was to maximize the life of the core tube and hence it was important to determine the settings of the parameters which yields maximum life of core tubes. The optimal settings was determined as follows:

Weld speed – High level (2 rev/sec)
Ramp out – High level (2 sec)
Ramp in – High level (1.5 sec)

The engineering team concluded that trial condition 4 (Table 9.18) gives the maximum core tube life. However, the desired value of the core tube was at least 8.5 million cycles. The above study conducted by the engineering team did not reveal any significant improvement to the process under investigation. Therefore a second case study was proposed with the aim of achieving better and satisfactory results.

Company's second attempt to use designed experiments

The second attempt was made with the assistance of author's skills and expertise in the area of study. A fishbone diagram (Figure 9.22) was constructed to identify the process parameters which influence the life of the core tubes. Twelve process parameters were initially thought to have some impact

Table 9.19 Effects of process parameters on core tube life

Process parameter	Average response at level 1	Average response at level 2	Effect
Weld speed	2.08	5.865	3.785
Ramp out	3.36	4.585	1.225
Ramp in	3.52	4.425	0.905

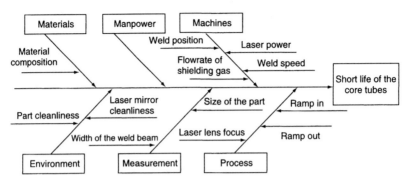

Figure 9.22 Fishbone analysis of the problem.

on the life. Further to a number of iterations, it was decided to select 5 out of 12 process parameters. Table 9.20 lists the process parameters along with their ranges of settings. The ranges of these parameter settings were determined after a thorough brainstorming with people from design, manufacturing, quality and shop-floor.

The following objectives were set by the company for the second round of experimentation. The objectives were determined by the team members and these were:

- to identify the laser welding process parameters which affect the mean fatigue life of core tubes
- to identify the process parameters which influence variability in life
- to determine the optimal settings of the process parameters which gives maximum life with minimum variability

For the second round of experimentation, the team has decided to study the following interactions.

1. $C \times D$
2. $A \times C$
3. $A \times D$

Table 9.20 List of process parameters and their ranges used for the second experiment

Process parameters	Label	Units	Low level	High level
Weld speed	A	Rev/sec	1.5	2.2
Ramp in	B	Sec	1.0	2.0
Ramp out	C	Sec	2.0	3.0
Laser power	D	Watts	950	1100
Lens focus	E	–	Position 1	Position 2

Choice of experimental layout for the experiment

For the second experiment, five main effects and three interactions were of interest to the team. The number of degrees of freedom for studying five main effects and three interactions (each parameter at 2-levels) is equal to eight. The best possible design matrix or experimental layout for this experiment was a $2^{(5-1)}$ fractional factorial experiment. This means that both main and interactions could be studied independently. The resolution of this design is V (i.e. main effects are clear of confoundings with two-way interactions and two-way interactions are free of confoundings with other two-way interactions). The following part explains the design generator and the confounding pattern of the design:

Design generator: $E = ABCD$
Defining relationship $= ABCDE$
Confounding pattern: $A = BCDE$, $B = ACDE$, $C = ABDE$, $D = ABCE$, $E = ABCD$, $AB = CDE$, $AC = BDE$, $AD = BCE$, $AE = BCD$, $BC = ADE$, $BD = ACE$, $BE = ACD$, $CD = ABC$, $CE = ABD$, $DE = ABC$

Table 9.21 displays the results of the second experiment with response values. Each experimental design point was replicated twice to increase the precision of the experiment. Moreover, the trial condition was also randomized to minimize the effect of bias induced into the experiment.

Statistical analysis and interpretation

In order to meet the objectives set at the outset of the project, it was important to perform statistical analysis of data generated from the experiment. If the

Table 9.21 Experimental layout and the response values for the experiment

Standard order	Weld speed	Ramp in	Ramp out	Laser power	Lens focus	Fatigue life (million cycles)
1 (7)	1.50	1.0	2.0	950	2.0	4.8, 1.3
2 (3)	2.20	1.0	2.0	950	1.0	6.3, 5.5
3 (10)	1.50	2.0	2.0	950	1.0	5.6, 4.8
4 (2)	2.20	2.0	2.0	950	2.0	9.0, 5.6
5 (15)	1.50	1.0	3.0	950	1.0	1.6, 2.9
6 (1)	2.20	1.0	3.0	950	2.0	8.4, 11.5
7 (9)	1.50	2.0	3.0	950	2.0	0.8, 4.1
8 (4)	2.20	2.0	3.0	950	1.0	8.3, 8.1
9 (14)	1.50	1.0	2.0	1100	1.0	2.0, 2.8
10 (5)	2.20	1.0	2.0	1100	2.0	4.8, 5.1
11 (12)	1.50	2.0	2.0	1100	2.0	4.7, 1.0
12 (8)	2.20	2.0	2.0	1100	1.0	5.0, 3.7
13 (16)	1.50	1.0	3.0	1100	2.0	4.6, 4.4
14 (6)	2.20	1.0	3.0	1100	1.0	8.0, 8.4
15 (11)	1.50	2.0	3.0	1100	1.0	5.0, 5.2
16 (13)	2.20	2.0	3.0	1100	2.0	10.8, 8.2

experiment was planned, designed, conducted and analysed correctly, then statistical analysis would provide sound and valid conclusions. The first step was to estimate the main and interaction effects of interest. Table 9.22 presents the table of effects and regression coefficients.

The identification of active and real effects is obtained with the help of Pareto and main effect plots. Figures 9.23 and 9.24 present main effect and Pareto plots. The figures indicate that two main effects (WS and RO) and two interaction effects (WS × RO) and (RO × LP) are found statistically significant at 5 per cent significance level. Here significance level is the risk of saying that a factor effect or interaction is significant when in fact it is not. The main effect and Pareto plots indicate that weld speed is the most active factor effect, followed by ramp out. The interaction between ramp out and laser power is shown in Figure 9.25. The interaction plot shows that life increases when the laser power is at high level and ramp out at high level.

It is quite interesting to note that although laser power on its own has very little impact on the life of core tubes, its effect on life is dependent on ramp out (Figure 9.25). In order to observe the effect of three factors on the mean

Table 9.22 Table of effects and regression coefficients

Term	Effect	Coefficient
A (WS)	3.819	1.595
B (RI)	0.469	0.235
C (RO)	1.769	0.885
D (LP)	-0.306	-0.153
E (LF)	0.369	0.185
A × C (WS × RO)	1.569	0.785
A × D (WS × LP)	-0.781	-0.391
C × D (RO × LP)	1.419	0.709

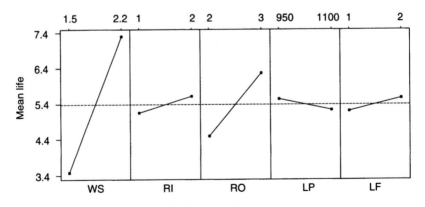

Figure 9.23 Main effects plot for the experiment.

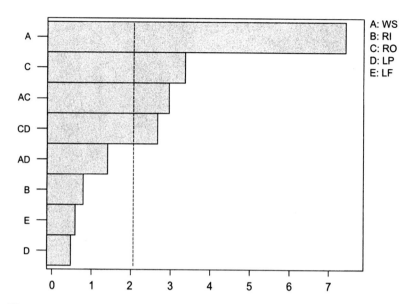

Figure 9.24 Pareto plot of effects affecting mean life.

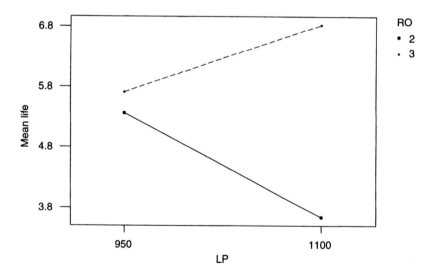

Figure 9.25 Interaction plot – ramp out × laser power.

life of core tubes, a cube plot is constructed (Figure 9.26). It is quite apparent in the cube plot that high level of weld speed yields a higher life. Similarly, it is fair to say that life increases with increase in ramp out.

The next step in the analysis was to identify the factors which influence fatigue life variability. To analyse variability SD was calculated at each experimental

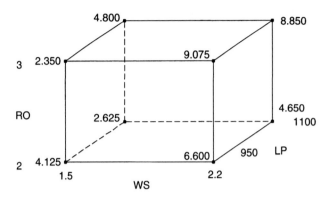

Figure 9.26 Cube plot of factors with mean life of core tubes.

design point. As log(SD) values tend to be normally distributed, a log transformation on SD values was essential. Table 9.23 displays the log(SD) values corresponding to each experimental trial condition. Due to insufficient degrees of freedom for the error term, it was decided to pool those effects with low magnitude. The Pareto chart (Figure 9.27) shows that the main effects lens position and laser power are significant at 5 per cent significance level. Similarly, it was also found that the interactions between lens focus and ramp in and laser power and ramp in were significant. Similar results can be obtained using analytical tools such as ANOVA (Analysis of Variance). For more information on the ANOVA, the readers are encouraged to refer to Montgomery's book (Design and Analysis of Experiments). Having identified the process parameters which influence the

Table 9.23 Table of log(SD) values

Trial number	log(SD)
1	0.394
2	−0.247
3	−0.247
4	0.381
5	−0.037
6	0.341
7	0.368
8	−0.851
9	−0.247
10	−0.674
11	0.418
12	−0.037
13	−0.851
14	−0.548
15	−0.851
16	0.264

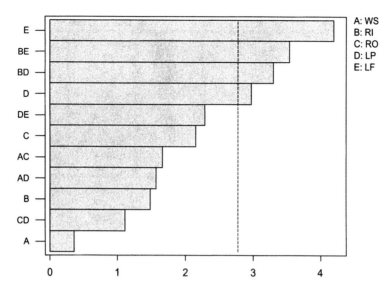

A: WS
B: RI
C: RO
D: LP
E: LF

Figure 9.27 Pareto plot of effects influencing variability.

mean and variability, the next stage was to determine the optimal process parameter settings that will maximize the core tube life with minimum variability.

Determination of the optimal process parameter settings

The selection of optimal settings of the process parameters depends a great
deal on the objectives to be achieved from the experiment and the nature of
the problem to be tackled. For the present study, the engineering team within
the company wants to discover the settings of the key process parameters that
will not only maximize the core tube mean life but also reduce variability in
core tube life so that more consistent and reliable products can be produced by
the manufacturer.

To identify the process parameter settings which maximizes the life, it was
important to select the best levels of those parameters which yield maximum
core tube life. This information can be easily generated from the main effects
plot (Figure 9.23). The interaction plot between ramp out (C) and laser
power (D) suggests that (Figure 9.25), the core tube life is maximum when
the laser power is set at its high level. Therefore, the optimal settings for
maximizing the core tube life is:

Weld speed (A)	Level 2 (2.2 rev/sec)
Ramp out (C)	Level 2 (3.0 sec)
Laser power (D)	Level 2 (1100 W)

In essence, the maximum core tube life was achieved only when all the above
process parameters were kept at high levels.

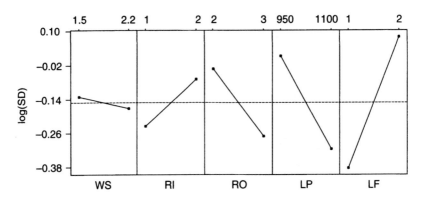

Figure 9.28 Main effects plot on variability (log(SD)).

In order to determine the best levels of process parameters which yield minimum variability, it was decided to construct a main effects plot on variability (using log(SD) as the response of interest). Figure 9.28 presents the main effects plot of process parameters for variability (log(SD) as the response).

The optimal settings for the significant process parameters which influence variability in core tube life is:

Ramp in (B)   Level 1 sec.
Laser power (D)   Level 2 (1100 W)
Lens focus (E)   Level 1 (position 1)

As there was no tradeoff in the levels of the process parameters, the final settings was determined by combining the above two. The final optimal condition is therefore given by:

Weld speed (A)   Level 2 (2.2 rev/sec)
Ramp in (B)   Level 1 sec
Ramp out (C)   Level 2 (3.0 sec)
Laser power (D)   Level 2 (1100 W)
Lens focus (E)   Level 1 (position 1)

Confirmation trials

Confirmation trials were performed in order to verify the results of the analysis. Five samples were produced at the optimal condition of the process. The mean life of the core tubes and tube life variance were 10.25 and 0.551, as opposed to 6.75 and 1.6 at the normal production settings in the company. This has shown an improvement of over 50 per cent in the life of the core tubes and a 65 per cent reduction in core tube life variability.

Significance of the study

Due to the significant reduction in process variability, the costs due to poor quality such as scrap, rework, replacement, re-test, etc. have reduced by over 20 per cent. This shows a dramatic improvement in the performance of the process and thereby more consistent and high quality core tubes could be produced using the optimized process. The engineering team within the company is now well aware of the do's and don'ts of experimental design. Moreover, the awareness that has been established within the organization about DOE has built confidence among the engineers and front-line workers in other areas facing similar difficulties. The author believes that it is important to teach a case study of this nature in order to learn the common pitfalls while applying DOE to a specific problem. The experiment also helped the engineering team within the company to understand the fundamental mistakes they make and indeed, the key features of making an industrial experiment a successful event.

9.2.8 Optimization of a spot welding process using Design of Experiments

This case study presents the application of DOE to a spot welding process in order to discover the key process parameters which influence the tensile strength of welded joints. Spot welding is the most commonly used form of resistance welding. The metal to be joined is placed between two electrodes, pressure applied and current turned on. The electrodes pass electric current through the work pieces. As the welding current is passed through the material via the electrodes, heat is generated, mainly in the material at the interface between the sheets. As time progresses, the heating effect creates a molten pool at the joint interface which is contained by the pressure at the electrode tip. Once the welding current is switched off, the molten pool cools under the continued pressure of the electrodes to produce a weld nugget.

The heat generated depends on the electrical resistance and thermal conductivity of the metal, and the time that the current is applied. The electrodes are held under a controlled pressure or force during the welding process. The amount of pressure affects the resistance across the interfaces between the work pieces and the electrodes. If the applied pressure is too low, weld splash (a common defect in spot resistance welding) may occur.

There are three stages to the welding cycle: squeeze time, weld time and hold time. The squeeze time is from when the pressure is applied until the current is turned on. The weld time is the duration of the current flow. If the weld current is high, it might again lead to weld splash. The hold time is the time which the metal is held together after the current is stopped.

As part of initial investigation and experiments were not performed before, the engineers within the company were more interested to understand the process itself. This understanding involved the key welding process parameters which affect the mean strength of the weld and also the process parameters which affect the variability in weld strength.

The following objectives therefore were set by a team of people within the company consisting of quality improvement engineers, process manager, two operators, production engineer and a DOE facilitator, who is an expert in the subject-matter. The objectives of the experiment were:

1. to identify the key welding process parameters which influence the strength of the weld
2. to identify the key welding process parameters which influence variability in weld strength.

Table 9.24 presents the list of process parameters along with their levels used for the experiment. As part of initial investigation, it was decided to study the process parameters at 2-levels. Owing to the non-disclosure agreement between the company and the author, certain information relating to the case study (process parameters, levels and original data) cannot be revealed. However, the data has not been manipulated or modified as a consequence of this agreement.

Interactions of interest

Further to a thorough brainstorming session, the team has identified the following interactions of interest.

(a) $A \times B$
(b) $B \times D$
(c) $C \times D$
(d) $D \times E$

The quality characteristic of interest for this study was weld strength measured in kg. Having identified the quality characteristic and the list of process parameters, the next step was to select an appropriate design matrix for the experiment. The design matrix shows all the possible combinations of process parameters at their respective levels. The choice of design matrix or experimental layout is based on the degrees of freedom required for studying the main and interaction effects. The total degrees of freedom required for

Table 9.24 List of process parameters used for the experiment

Process parameter	Label	Low level setting	High level setting
Stroke distance	A	−1	1
Weld time	B	−1	1
Electrode diameter	C	−1	1
Welding current	D	−1	1
Electrode pressure	E	−1	1

studying five main effects and four interaction effects is equal to nine. A $2^{(5-1)}$ fractional factorial design was selected to study all the main and interaction effects stated above. The degrees of freedom associated with this design is 15 (i.e. $16 - 1$).

In order to minimize the effect of noise factors induced into the experiment, each trial condition was randomized. Randomization is a process of performing experimental trials in a random order in which they are logically listed. The idea is to evenly distribute the effect of noise across (those which are difficult to control or expensive to control under standard production conditions) the total number of experimental trials. Moreover, each design point was replicated five times to improve the efficiency of experimentation. The purpose of replication is to capture variation due to machine set up, operator error, etc. Moreover, replications generally provide estimates of error variability for the factors (or process parameters). Table 9.25 illustrates the results of the experiment.

Statistical analysis of experimental results

Statistical analysis and interpretation of results are imperative steps for DOE to meet the objectives of the experiment. A well-planned and designed experiment will provide effective and statistically valid conclusions. The first step in the analysis was to identify the factors and interactions which influence the mean weld strength. The results of the analysis are shown in Figure 9.29. The Pareto plot (Figure 9.29) shows that main effects D (welding current) and E (electrode pressure) have significant influence on mean weld strength. Moreover, two interactions A × B (stroke distance × weld time) and B × D (weld time × weld-welding current) are also found to be statistically significant. Main effects A, C and B did not have any influence on the mean weld strength.

Table 9.25 Results of the experiment

Run	A	B	C	D	E	Mean weld strength
1	−1	−1	−1	−1	−1	5.4
2	1	−1	−1	−1	1	20.4
3	−1	1	−1	−1	1	243.0
4	1	1	−1	−1	−1	109.0
5	−1	−1	1	−1	−1	48
6	1	−1	1	−1	1	104
7	−1	1	1	−1	1	23.6
8	1	1	1	−1	−1	3.40
9	−1	−1	−1	1	−1	763
10	1	−1	−1	1	1	750
11	−1	1	−1	1	1	553
12	1	1	−1	1	−1	279
13	−1	−1	1	1	−1	462
14	1	−1	1	1	1	610
15	−1	1	1	1	1	747
16	1	1	1	1	−1	576

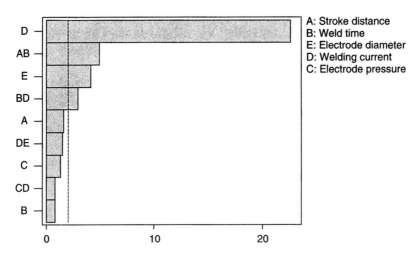

Figure 9.29 Pareto plot of main and interaction effects from the experiment.

In order to analyse the strength of the interaction among the process parameters stroke distance, weld time and welding current, it was decided to construct interaction graphs (Figures 9.30 and 9.31).

Figure 9.30 shows that high weld time and low stroke distance yield highest weld strength. Moreover, high weld time and high stroke distance yield lowest weld strength. Similarly, Figure 9.31 indicates that high welding current and low weld time yield highest weld strength. Here there is a tradeoff in the selection of factor levels for weld time. However further studies showed that high weld time and high welding current combination produces the highest weld strength.

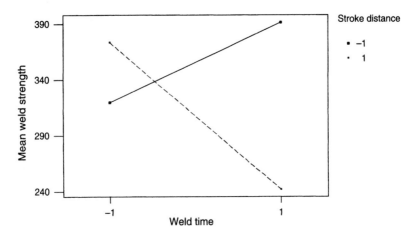

Figure 9.30 Interaction graph for weld time and stroke distance.

Figure 9.31 Interaction graph for welding current and weld time.

One of the assumptions experimenters generally make in the analysis part is that the data come from a normal population. In order to verify that the data follow a normal distribution, it was decided to construct a NPP of residuals (residual = observed value − predicted value). Figure 9.32 presents a NPP of residuals which clearly indicates that all the points on the plot come close to form a straight line. This implies that the data are fairly normal.

The next step in the analysis was to identify the key process parameters which affect variability in weld strength. To analyse variability, SD was calculated at each experimental trial condition. As ln(SD) values tend to be normally distributed, a log transformation was carried out on the data. The results are shown in Table 9.26.

In order to identify which of the factors or interactions have a significant impact on variability in weld strength, it was decided to construct a Pareto plot (Figure 9.33). The graph shows that only welding current has a significant

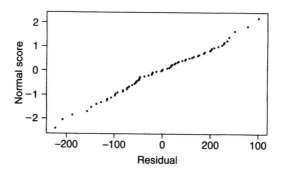

Figure 9.32 Normal probability plot of residuals.

Table 9.26 ln(SD) values from the experiment

Trial number	ln(SD)
1	1.086
2	2.961
3	3.642
4	3.713
5	4.008
6	3.481
7	3.379
8	1.329
9	4.011
10	3.379
11	3.931
12	4.937
13	3.646
14	3.560
15	4.000
16	4.070

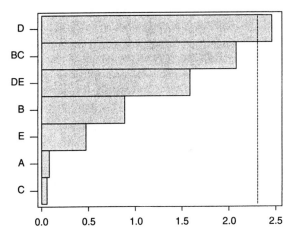

Figure 9.33 Pareto plot of effects on variability in weld strength.

impact on variability in the strength of the weld. In order to generate adequate degrees of freedom for analysing variability, pooling was performed (by combining the degrees of freedom associated with those effects which are comparatively low in magnitude). In order to support the procedure of pooling, a NPP of effects was also constructed. It is interesting to note that variability in the strength was minimum when welding current was set at low level of setting. As there was a tradeoff in one of the factor levels (factor D), it was decided to perform the loss-function analysis promoted by Dr Taguchi.

Loss-function analysis for Larger-the-Better (LTB) characteristics

This analysis is used when there is a tradeoff in the selection of process parameter levels. As the performance characteristic of interest in this case is strength of the weld, it was decided to perform the loss-function analysis for LTB performance characteristics. The average loss function for LTB quality characteristic is given by:

$$L = k\left[\frac{1}{\bar{y}^2}\right]\left\{1 + \left(\frac{3\ SD^2}{\bar{y}^2}\right)\right\} \tag{9.2}$$

where k = cost constant or quality loss coefficient, $\bar{y}$ = mean performance characteristic (i.e. mean strength), SD = standard deviation in the strength of the weld corresponding to each trial condition and L = average loss associated with the performance characteristic per trial condition.

Equation (9.2) is applied to all 16 trial conditions. It was found that trial condition 10 yields minimum loss. For trial condition 10, factor D was set at high level and therefore high level setting for D was chosen for the model development and prediction of weld strength.

Significance of the study

The purpose of this paper is to illustrate an application of DOE to a spot welding process. The objectives of the experiment in this study were twofold. The first objective was to identify the critical welding process parameters which influence the strength of the weld. The second objective was to identify the process parameters which affect variability in the weld strength. A trade off in one of the factor levels (factor D) was observed. This problem was rectified with the use of Taguchi's loss function analysis. The strength of the weld has been increased by around 25 per cent. The next phase of the research is to perform more advanced methods such as Response Surface Methodology (RSM) by adding center points and axial points to the current design. The results of the experiment have stimulated the engineering team within the company to extend the applications of DOE in other core processes for performance improvement and variability reduction activities.

9.3 Summary

This chapter presents eight experiments to illustrate the power of DOE in real life situations. Each study clearly presents the nature of the problem or objective(s) of the experiment, experimental layout chosen for the experiment, analysis and interpretation of data using powerful graphical tools generated by Minitab software system. The case studies presented in the book would stimulate engineers in manufacturing companies to use DOE as a powerful technique for tackling process or product quality related problems.

References

Antony, J. (1999). *Improving the wire bonding process quality using statistically designed experiments*, 30, 161–168.

Bullington, R.G. et al. (1993). Improvement of an Industrial thermostat using designed experiments, *Journal of Quality Technology*, 25(4), 262–270.

Crafer, R.C. and Oakley, P.J. (1981). Design principles of high power carbon dioxide lasers, *The Welding Institute Research Institute Bulletin*, pp. 276–279.

Dodson, B. and Nolan, D. (1999). *Reliability Engineering Handbook*. Tucson, AZ, U.S., QA Publishing.

Green, T.J. and Launsby, R.G. (1995). Using DOE to reduce costs and improve the quality of microelectronic manufacturing processes, *International Journal of Microcircuits and Electronic Packaging*, 18(3), 290–296.

Hamada, M. (1995). Using Statistically designed experiments to improve reliability and to achieve robust reliability, *IEEE Transactions on Reliability*, 44(2), 206–215.

Irving, B. (1996). Search goes for the perfect Resistance Welding Control, *Welding Journal*, 75(1), 63–68.

Logothetis, N. and Wynn, H.P. (1989). Quality through Design – Experimental Design, *Off-line Quality Control and Taguchi Contributions*, Oxford, UK, Oxford Science Publications.

Minitab user's Guide (2000). *Data Analysis and Quality tools*. Release 13. UK, Minitab Inc..

Montgomery, D.C. (1992). The use of statistical process control and Design of Experiments in product and process improvement, *IIE Transactions*, 24(5), 4–17.

Sirvanci, M.B. and Durmaz, M. (1993). Variation reduction by the use of designed experiments, *Quality Engineering*, 5(4), 611–618.

William, G.W. (1990). Experimental design: Robustness and power issues, *ASQ Congress Transactions*, pp. 1051–1056.

Index

Lightning Source UK Ltd.
Milton Keynes UK
UKOW031809300112

186342UK00002B/24/P